THE

ADAPTIVE MILITARY

with a new introduction by the editor
SECOND EDITION

—— THE ——
ADAPTIVE
MILITARY

Armed Forces in a Turbulent World

—— *edited by* ——
James Burk

Transaction Publishers
New Brunswick (U.S.A.) and London (U.K.)

Second printing 2002

Copyright © 1998 by Transaction Publishers, New Brunswick, New Jersey.

Library of Congress Catalog Number: 98-28518
ISBN: 0-7658-0472-7
Printed in the United States of America

Library of Congress Cataloging-in-Publication Data

The adaptive military : armed forces in a turbulent world / James Burk,
 editor ; with a new introduction by the editor. — 2nd ed.
     p. cm.
 Rev. ed. of: Military in new times. 1994.
 Includes bibliographical references and index.
 ISBN 0-7658-0472-7 (alk. paper)
 1. Armed Forces. 2. Military policy. 3. World politics—1989– I. Burk,
James, 1948–   . II. Military in new times.
UA11.A33   1998
355'.0335—dc21   98-28518
                                                            CIP

Contents

List of Tables

Acknowledgments

This book perhaps more than most has incurred many debts in its making. Professor Joseph G. Dawson III, who is director of the Military Studies Institute at Texas A&M University, first suggested that I undertake this project. After I agreed, he strongly supported my work, and the work of the other contributors, with his prudent counsel, unfailing good humor, and the Institute's material resources, all of which we needed to complete our work. Under the auspices of the Military Studies Institute, the contributors were able to meet in a symposium held at Texas A&M in April 1992, there to explore the ideas that are reported in full in this volume. I am indebted also to R.J.Q. Adams, who chairs the Military Studies Institute Advisory Council, and to James Bradford, Larry D. Hill, Brian Linn, and Alex Mintz, who are members of the Council. They provided much sound advice and welcome encouragement. Laura Ampol Hall worked indefatigably to help when I needed her most, looking after the large number of details invariably associated with an undertaking such as this one, to include doing her fair share of word processing. I am grateful for her help. Finally, Ensign Theodore M. Burk, U.S. Navy, and Patricia G. Burk, who have over the years helped me in many ways, helped here too in ways they best know.

Acknowledgments (1998 Edition)

This book has not been revised since its original publication, except to correct a few minor errors. It has been enlarged with a new introduction to explain why a new edition may be helpful. I have been encouraged in this enterprise by Michael Anderson, Christopher Dandeker, Morton Ender, David Segal, and Don Snider and thank them for their encouragement. My thanks also to Irving Louis Horowitz and Mary E. Curtis for their help and hard work in bringing this edition to light.

James Burk

Introduction, 1998: Ten Years of New Times

James Burk

Nearly ten years have passed since the end of the Cold War and with them also the euphoria that initially accompanied release from that bipolar conflict. The Cold War consumed the attention and energy of much of the world for two generations. When it ended, many hoped it signified enhanced prospects for a more stable world, a world in which the threat of general war would recede under the influence of expanding prosperity, a world in which genuine peace would be underwritten by the cooperation of superpowers and by the establishment of a new wave of market–based, liberal democracies.

These hopes were not baseless. A "velvet revolution" in Eastern Europe had replaced communist autocracy with more liberal economic and political societies. These societies reinforced a trend toward democratization that was (and is still) evident in many places around the globe. Following the collapse of the Soviet Union, even Russia seemed ready to jettison its deeply ingrained authoritarian political culture and to embrace market capitalism.[1] These favorable political developments occurred within the context of an unprecedented economic expansion. While the benefits of this expansion were unevenly distributed around the globe and within particular societies, prosperity was widespread. Confidence that the expansion will continue has been shaken but not eroded by the recent financial crises in Asia. Pacific military trends may be added to these hopeful political and economic developments. The most powerful countries of the world, as measured by military might, have reduced the size of their armed forces and the flow of resources in support of military purposes. More important perhaps, and despite continuing tragic outbreaks of collective violence, there is no evidence of any rise in the incidence of wars between or within states.

Nevertheless, whoever thought the post–Cold War era would bring an "end to history" or establish a "new world order" has been disappointed. The period has been marked by turbulence, uncertainty, and challenge. Dealing with rogue states like Iraq and North Korea has shown that checking proliferation of nuclear weapons and other means of mass destruction is, if anything, more troublesome now than it was during the Cold War.[2] The experience of war in the wake of failed states in Somalia, Bosnia, and elsewhere has raised doubts about whether (and, if so, how) multinational peacekeeping operations can ameliorate the social chaos and human misery brought on by the collapse of states. There is at best an uneasy consensus about the purpose or future of collective security arrangements, brought into sharp focus by debates over the wisdom of expanding the NATO alliance to include new states in Eastern Europe.[3] And hopes for expanding the "zone of peace" through increased democratization are dimmed by assertions that the values of liberal democracy and respect for human rights are "impositions" of Western culture, incompatible with the values of non–Western countries, especially in Asia, and so are not appropriate for all countries to adopt.[4] Even defenders of democracy wonder whether it may not in many cases be destabilizing for international organizations to require states, as a condition for receiving aid, to adopt the democratic practice of "free elections" to allocate political power.

In short, after ten years of "new times," no simple judgment of the state of world politics or, specifically of world security—whether positive or negative—is likely to be correct. Qualifications are required. One purpose of this volume is to identify as clearly as possible what some of those qualifications may be, in particular as they relate to military security policies and practice, and to explain why they are necessary.

The chapters which follow were originally written for a conference held in 1992, much closer to the time when hopes for a new era were higher than they are today. To some degree they reflect the mood of those times. Indeed, sympathetic reviewers of the first edition questioned what they thought were Manchesterian assumptions underlying many of the analyses and they criticized our relative neglect of continuing problems posed by nuclear proliferation and the threat of general war.[5] But, after reviewing the essays in preparation for this new edition, I am struck rather by their realistic appreciation for the difficulties posed by a world politics no longer organized around a bipolar superpower conflict and by the continuing relevance of their main themes and guiding hypotheses, especially for those who would understand the multiple factors affecting the mission, organization, and use of armed forces by the major NATO powers. The central argument they pursue is that the "new times" of turbulence, following the Cold War, have been shaped by long–term trends that operated throughout the Cold War and sometimes longer.[6] These trends encouraged an increase in the number and complexity of threats that the armed forces of the developed world are supposed to meet. At the same time, in certain ways, they decreased the stock of social and political capital available to meet them.

In this introductory essay, I briefly sketch how the different chapters contribute to the development of this central theme. Their contributions do not by any means speak with a single voice. The contributors differ in their assessments about the current prospects for peace and ways to maintain military security, and there is no sense in trying prematurely to impose consensus. My aim rather is to highlight the continuing relevance of what they say by locating their arguments in the context of current unresolved debates. Their answers may vary, but the major questions they address are as critical now as they were ten years ago. Four questions are especially relevant. First, how has the state system been transformed from what it was during the era of the world wars to what it is today? Why talk about "new times" in world politics? Second, what are the consequences of this transformation for the organization of collective violence? What should we include within the realm of security threats? Third, how have the armed forces of advanced industrial nations tried to adapt themselves to these new times? And, finally, what are the present prospects for using armed forces to achieve a more peaceful world?

The Idea of "New Times"

Each moment in history, of course, is obviously and trivially "new." Each presents a never–before–seen constellation of material resources and restraints, of hopes for justice and helplessness in the face of oppression, and of unexpected dreams and old customs, traditions, and beliefs. Taken in this sense, claims that the end of the Cold War opened a new era in world politics are impossible to dispute. Yet, unless we subscribe, unwisely, to a deterministic theory that would resolve all our questions by asserting that some abstract idea is the force directing history—racial superiority, reason, class struggle, free market liberalism, etc.; there are many alternatives to choose from—it is important to clarify just what is new about the present, to say how the present differs from the past when so much of the past remains an active force in our lives.[7] In our case, it is important to know how it differs in ways that affect understanding of what current security issues are and how militaries prepare to meet them.

Perhaps the simplest thing to say is that, very soon after the Cold War began, a shared interpretative framework developed that enabled us to comprehend the nature of the conflict and the organization of international relations around it. This framework was built on a foundation of clear and powerful symbols—Cold War; iron curtain; East and West; open versus closed societies; the free world; the communist bloc; first, second, and third world countries, etc. Our use of these terms led us to see world politics and international security issues in a particular way. Yet with the Cold War's end, this language for thinking about the world was no longer relevant; the vision it supported was moribund, without meaning. Still, it had been so long since any other language was employed, no other interpretative framework was ready to use and none was able to make as much sense of the world as the Cold War rhetoric seemed to do.[8] Even after a decade, no powerful metaphors have been coined, as happened in the Cold War by the late 1940s, to describe the period we have entered. (The phrase "post–Cold War" is often used, but it is obviously derivative and hardly counts as a powerful or useful political symbol.) In short, we may believe that these times are new (qualitatively different from the Cold War) if only because we no longer possess a common coherent framework within which we can argue about the meaning of ethnic cleansing in Bosnia, secessionist movements in Eastern Europe; slaughter in Algeria, Rwanda, and Burundi; and Chinese saber rattling over Taiwan.

There is an explanation for this sharp contrast between the quickly formed and rich rhetorical framework that arose during the Cold War and the relative absence of such a framework to describe the last decade. It has to do with the character of the international order itself.

The bipolar struggle between the United States and the Soviet Union was unique in many respects. Consider just three. It was the first major conflict among powers wielding weapons of mass destruction the unrestrained use of which surely would have had horrifying consequences for the civilized world, to say the least. Perhaps for this reason, it was a period of conflict in which the two major protagonists did not often risk direct military confrontation, but fought indirectly and fitfully through proxies, who often possessed independent minds and took advantage of the superpowers to further purposes of their own.[9] And it ended without shots being fired, the contest turning finally on the relative strength of the economies to endure the burden of an arms race and preparation for war while at the same time having to supply the material needs of their own populations. For all its uniqueness, however, it was possible to understand that conflict as a contest for power and influence between two strong and sovereign states, indeed the leading powers in a system of states whose competitive relations defined world politics. While varied in their causes and conduct, such contests have been the major form of global conflict since the modern state system was established by the Peace of Westphalia in 1648. The architects of the Cold War's interpretative framework could build a rhetorical structure from materials that were familiar from over three centuries of use. The security landscape of the Cold War was unique in scale and consequence, perhaps, but it was no terra incognita.

As James Rosenau argues, it is no longer possible to comprehend world politics in terms limited to a Westphalian system of international relations among sovereign nation–states, as if these were the only important actors on the scene.[10] State sovereignty (with all that implies about the state's power to use armed force in pursuit of its interests) is challenged now on many fronts. Over the course of this century, the number and variety of states in the world have vastly multiplied, increasing over sixfold since the beginning of the century when there were fewer than thirty states in all. While many of the new states are weak, their very existence represents a loosening of the grip of European colonial rule and a reduction in the relative autonomy of Western states. These trends have been reinforced and broadened by the introduction of many new nonstate institutions—for example, the United Nations, the Red Cross, the International Monetary Fund, Greenpeace, Habitat for Humanity, Human Rights Watch, Amnesty International, and multinational business corporations, to name a few—onto the stage of world politics where they have a substantive role to play. (Even this extensive list ignores the growing forcefulness of ethnic groups, religious fundamentalists, and the international organization of criminal and terrorist activities, all three of which have also altered the character of world politics.) Unrestrained by the responsibilities of sovereignty, this panoply of nonstate actors (sometimes, including even individuals) is sometimes powerful enough to challenge the authority of states and to affect the contours of world conflicts.[11] States are less certain of their authority and much less certain of the utility of using armed force to make war against other states or to ensure order and stability at home.

On such a cluttered stage, with so many actors and so many scripts, it is difficult to imagine a satisfactory narrative of world politics that draws exclusively upon the prevailing realist (or neorealist) account of international relations. The realist narrative presumes that world politics is an anarchy driven by conflicts among sovereign states anxious to defend their territory and interests against encroachment by other states. In this anarchy, interstate cooperation is possible, but it cannot be counted on. The assumption is always that the other is a threat and that, when calculating threats, the other's motives (however benign) matter less than capabilities. To survive, armed vigilance is required. Stability, if it exists, results from the balance of (ultimately coercive) power, which the prudent ruler assumes to be temporary instead of permanent. This account is not wrong.[12] Certainly, even now, states wield enormous power and are the principal actors in world politics. But the account is incomplete. Increasingly, since the end of the Second World War, it requires augmentation. While interstate rivalries and conflicts persist, it is critical to note that, in many places, traditional wars between states, while still being fought, are no longer the major form of armed conflict. Moreover, in a world where many actors challenge state authority, states have formed cooperative relationships, especially in Western Europe and North America and other areas where there is a stable "zone of peace." They act to pursue common interests which they can only achieve together, developing collective identities above the state, a new basis "for feelings of solidarity, community, and loyalty."[13] This development is not contemplated by the realist account. Where stability is lacking and collective violence thrives, it is usually not the result of unresolved interstate conflicts, but rather of the inability of weak states to maintain order and popular support within their own realms.[14] In short, it is when states disappear that world politics turn violent.

What narratives, if not the realist one, will best explain the new situation? That remains an unanswered question. But it is no wonder (nor cause for lament) that we have no ready answer as we did for the Cold War. The international order today is qualitatively different from what it was through the Cold War and long before. State sovereignty—the relative autonomy of states to act on the world stage—is much more variable (harder to take for granted) than it used to be. States are no longer as strongly sovereign in their own realm nor in the realm of international affairs as they once were—or thought themselves to be. Now transnational organizations tower over states and constrain what they can do and subnational groups challenge the legitimacy and sometimes even the relevance of states. How do these facts affect the security threats states face, their military organizations, and the prospects for peace?

Assessing Current Security Threats

There is no simple translation from the current turbulent and multifaceted organization of world politics to a definition of security threats. Nevertheless, the major security problems with which the post–Cold War world is concerned are by now well–known. They are, first, the problem of deterring the use and proliferation of weapons of mass destruction and, second, the problem of what Kalevi J. Holsti has called "wars of a third kind."

The problem of controlling weapons of mass destruction is not unique to the present; it was perhaps the central issue for Cold War strategic doctrine and lay at the heart of deterrence theory. Arguably, dealing with the problem requires that possession of these weapons should be concentrated in the fewest possible hands and maintained under strict and highly centralized bureaucratic command. This was difficult enough to do, however, during the Cold War when the United States and the Soviet Union held a virtual (never complete) monopoly over such weapons. The problem now is more complex. The technology for producing these weapons is more widely known and the means of delivering them are within the reach of many medium–sized states. Unhappily, deadly chemical and biological agents may be produced and unleashed even by relatively small, but committed terrorist organizations. Preventing proliferation of the weapons of mass destruction, in short, is a matter of high priority.[15] Permanently extending the Treaty of the Non–Proliferation of Nuclear Weapons was one important but only a partial step to curb the threat, if only because chemical and biological weapons need to be considered. United Nations experience with Iraq has shown that curbing proliferation requires heavy investments in intelligence gathering and verification regimes. Enforcing nonproliferation also requires heavy investments in counterforce capabilities able to neutralize a weapons stock that may be under an adversary's control. Failure to maintain and update Cold War deterrence efforts to control present threats exposes all states to unacceptable risks of destruction. Indeed, it has recently been argued that "the proliferation of nuclear, biological, and chemical (NBC) weapons is the main strategic threat facing the U.S. and its allies."[16]

No one knows whether failure of deterrence might trigger a general war. At the moment, fortunately, the likelihood of general war—the wars for which weapons of mass destruction were originally developed—remains low. There are no focal enemies and no set of major powers who seriously contemplate going to war with one another.[17] Are we justified therefore in supposing that general war is no longer a genuine threat? William R. Thompson is leery of that supposition. His contribution to this volume exposes the logical weaknesses of various theories (Rosenau's among them) that claim general wars are extinct, destined, like dinosaurs, never again to stride the earth.[18] He agrees with most observers that such war is unlikely in the foreseeable future, but thinks it unwise to project the present situation too far into the future. Arguments that general war will never return depend on the validity of enlightenment and whiggish beliefs that wars will disappear as Western values, institutions, and preferences become the global standard and thrive happily without resistance or reversal anywhere in the world. Stated starkly, such expectations seem naive and contrary to history. Thus Thompson urges an approach to military security policies that guards against resurrection of general war.[19] An effective regime of deterrence against weapons of mass destruction is critical to the success of that effort.

The second major security threat is the problem of "wars of a third kind." Wars of a third kind must be distinguished from either the highly ritualized or institutionalized wars that characterized interstate warfare during the eighteenth century or the industrialized total wars with which we became familiar early in this century. Wars of a third kind are usually intra– rather than interstate wars. They are protracted conflicts, with no declarations to mark their beginning and no peace settlements to mark their end. They are irregular wars, fought by people who wish to have a country of their own. There are no regular uniformed armies fighting for honors on a battlefield; there are no campaigns with set strategies or tactics; and there are few if any distinctions made between soldiers and civilians. War by people against people, not states against states, these conflicts are often bloody, especially for civilian populations. As we have seen too often, thousands are tossed from their homes, massacred, or raped, and those who are not often escape only by becoming refugees, with their flight usually upsetting domestic peace no matter where they go.[20] Such conflicts have become the predominant form of violent conflict over the last fifty years. Based on Holsti's count of 164 major armed conflicts from 1945 to 1995, 126 (or 77 percent) have been intrastate wars of this kind.[21]

The frequency of intrastate wars since 1945 is tied in part to the development of weapons of mass destruction. These weapons are so powerful that they have made it difficult to correlate traditional measures of military strength with strategies for military action. Major powers are reluctant to unleash the force at their disposal because they fear (quite rightly) that to do so without restraint would be self–defeating. Indeed, even conventional weapons are powerful enough that those who resist the state by violent means are unwilling to risk direct confrontations. They prefer to rely on tactics of indirect irregular or guerrilla warfare and on episodic acts of terrorism, counting on these to wear down the will and strength of their adversaries.[22] Many used these means in wars of national liberation waged to win independence from colonial powers following the Second World War. Consequently, we might hope that this is a transitory form of war, whose high incidence only reflects instabilities connected with the dismantling of the colonial empires.

But our hope would rest on shaky ground. First, the logic of wars for national liberation has been generalized over the course of this century (under the doctrine of self–determination) to underwrite ethnic, clan-based, and religious struggles for political autonomy all around the world, from Algeria to (what was) Zaire. Second, wars of a third kind thrive wherever states are weak or, even worse, have collapsed altogether as happened in Lebanon and Somalia. Unfortunately, many states are weak. State strength depends on a shared understanding of the state's existence, of the boundaries of the territory over which it rules, and of the legitimacy of its central institutions to exercise authority.[23] For a variety of reasons, the experience of colonial rule among them, states frequently fail to meet these conditions. Israel, for instance, is a relatively strong and successful democratic state. Yet it is weakened and divided by Palestinians who contest Israel's right to exist, its borders, and the authority of its ruling institutions and who wage irregular warfare against it. Israel's ability to respond to the Palestinian challenge, one way or another, is weakened because there are sharp and contradictory differences among various Jewish groups—say between secular and ultra-Orthodox Jews—over the mission the state, its "natural" borders, and the division of institutional authority. Of course, no state is perfectly strong. But weaker states are less able to command loyalty from their people or to extract resources needed to administer government effectively. They are more likely to resort to coercion as a substitute for the rule of law and more likely to invite acts of hostility against them. In some cases, this creates a revolutionary situation, as it did during the last days of the Shah in Iran. At worst, the state evaporates altogether, as happened in Somalia, and people are left to live in strife and lawlessness. More common are cases less extreme, where states do not collapse, but carry on at a low level of effectiveness. In such situations, they may become havens for (or allies with) outlaw groups, terrorists, drug dealers, and so on.

Violent conflicts within weak states pose diffuse threats difficult for major powers to assess.[24] They are threats which do not arise close to home or where major powers believe their "vital" interests lie. Western Europe, North America, and, with some qualifications, South America are regions of strong states, within which there are virtually no prospects for interstate war or indeed for wars of any kind. East Asia, perhaps, may be included in this category, despite concern over China's growing military power. High levels of economic interdependence and cooperative ventures to maintain economic well–being may make Asian leaders reluctant to engage in a military conflict that could destroy the bases of their own wealth.[25] Africa, the Middle East, and Eastern Europe, in contrast, define a region of weak states within which wars of a third kind, illegal drug running, massive immigration flows, terrorist attacks, and humanitarian disasters are most likely to occur. Examining this problem from the American perspective, Donald Snow highlights the consequences of this mismatch between national interests and threats.[26] During the Cold War, the threat of collective violence seemed greatest where "vital" U.S. interests were at stake. This alignment helped to clarify U.S. security policy and to build consensus for it. But now, where American "vital" interests are highest—in the Americas and Western Europe—the threat of violence is low, and where the threat of collective violence is highest—in Africa, say—American "vital" interests are not at stake. When interests and threats are out of alignment, an appropriate military security policy is harder to define and harder to justify. Political leaders have to explain why it is necessary to risk American lives for humanitarian disasters that occur far from home and have little direct "pay–off" for the nation. What is true for the U.S. in this regard is also true for other major powers.[27]

Adapting to Change

How have military organizations adapted to this new security environment? If there is a single answer to the question, it is that they are smaller than they used to be. This can be measured by spending. U.S. Army Lt. Gen. Patrick M. Hughes, director of the Defense Intelligence Agency, recently testified before the Senate Intelligence Committee that global defense spending is 40 percent less than it was ten years ago.[28] Defense spending by the United States follows the global pattern, despite its role as a leading power. Department of Defense outlays dropped from 6.0 percent of GDP in fiscal year 1987 to 3.4 percent of GDP in fiscal year 1996. The decrease represented a real cut in defense spending, not simply a failure of defense budgets to keep pace with economic growth, and was achieved in large part by reducing U.S. force size by one–third between fiscal years 1987 and 1997.[29] In Western Europe, countries that once relied heavily on conscription to raise military forces are moving almost in unison either to abolish it (as Belgium and the Netherlands have done and France is doing) or to lighten its burden. At the end of the Cold War, half of the male citizens in Western Europe between the ages of eighteen and thirty-two were conscripted for military service. By 1995, only one in three were called. Further declines are expected.[30] The problem is to know what these declines signify. One could say, simply, that they reflect and confirm the general assessment that major interstate conflicts are not likely in the foreseeable future. While true, the claim could be misleading in two ways. First, it might make it seem that downsizing followed naturally on the heels of a judgment that militaries suddenly have much less to do. Second, it glosses over a growing number of domestic controversies about what the military should be doing and how it should be related to the society on whose support it depends.

Functionally, in the current threat environment, the major NATO powers have more and more varied rather than fewer missions to perform. They retained the mission of preparing for and, if necessary, fighting large–scale wars. At the same time they have added responsibilities to conduct "operations other than war." These include a wide–ranging bundle of tasks that respond both to domestic security requirements and to threats posed by weak states and wars of the third kind, for example, strategic and traditional peacekeeping, protection against terrorist threats, intelligence gathering to curtail proliferation of weapons of mass destruction, control of immigration and refugee flows, humanitarian and disaster relief, and so on. With so much to do, why have armed forces been subject to such drastic budget cuts and reductions in force size?

Christopher Dandeker suggests that the reasons why have less to do with the end of the Cold War than with two broad secular trends affecting the exercise of public authority.[31] First, central governments have had to impose tighter controls over spending to stimulate development of national economies able to compete in a global market economy; large budget deficits and high tax burdens, which had been common, are now impossible to sustain.[32] Yet striving after greater fiscal discipline elevates the importance of market–based norms of efficiency in public budget decision making, no matter whether these are substantively appropriate. The result is to restrict the sphere within which military leaders may frame their judgments about force structure and performance according to the logic of their "objective" professional expertise. Economic logic has higher priority. Second, the influence of market norms is intensified against the background of the public's declining willingness uncritically to accept traditional claims to professional autonomy. In this context, professional soldiers, like professionals in other spheres of life, are required to justify not only their claims for resources, but also their policies and their practices, in terms of values that are not necessarily embedded in the technical requirements of their job. In the United States, for instance, federal courts have moved recently to overturn administrative and legal actions by military leaders in order to extend the protection of constitutional rights to military personnel. Traditionally, the court's have been reluctant to do this; instead they deferred to the idea that the military was a unique institution, exempt from the usual standards of review based on established constitutional norms.[33]

Sometimes these trends interact in ways that exacerbate the difficulties political and military leaders face as they try to adapt to the new security environment. If the public, for instance, insists that casualties in peacekeeping operations be kept to an absolute minimum, there is a cost to pay in developing (and training with) high–technology weapons that both protect personnel and deliver force effectively to accomplish the mission. Obviously, this is difficult to do with shrinking budgets and smaller forces. How then to proceed? Recent efforts at force restructuring try to leverage assets by relying increasingly on reserve forces to complement the active duty cadre. Even so, the armed forces of most major powers remain "overstretched" and find it difficult to maintain morale and efficiency. In Western Europe, meeting the challenge posed by these trends has encouraged development of transnational military organizations to take advantage of economies of scale and scope.[34]

There are also less obvious, but profound and paradoxical consequences to consider. Over twenty–five years ago, Morris Janowitz foresaw movements in this direction and expected that they would lead to a more politicized military, one that would mobilize as a political pressure group to defend its preferred policies and national budget shares.[35] Since the end of the Cold War, we have accumulating evidence that these expectations are being met. As Eliot Cohen recently observed, military officers, like doctors and lawyers, have

> shed the image of pure and apolitical expertise once ascribed to them.... Indeed, it is not uncommon for officers to describe themselves as a governmental interest group and to justify (if somewhat abashedly) their collective actions in such terms.[36]

This development is sufficiently troubling in the United States that students of the military have debated whether there is a "crisis" in civil–military relations that is unhealthy for the country.[37] The debate has raised a number of important questions. Not the least important one concerns the military's role in determining whether and how force should be used. In the current geopolitical context, it is often unclear how force may be used to achieve a given policy objective, once the obvious but crude objectives of conquest or defense against conquest are ruled out of play. There is room for debate and need for good advice. Yet, how much influence over this debate should be wielded by a military elite that sees itself as a political interest group? The question is given added piquance once we know that the political values of the American military elite are increasingly homogeneous and divergent from the political values held by members of other elite groups.[38] Under these circumstances, if the military elite is accorded great influence, whether direct or indirect, the country may pursue policies or pursue policies in a way that it would prefer not to do—assuming it was fully informed.

Obviously more than material resources are at stake. At issue is how the military and society are related. Charles Moskos and I believe that there is a deficit of social capital to support armed forces, despite survey data showing that the public has a relatively high level of confidence in the military as an institution and in its leaders.[39] The relationship is, after all, too complex to be captured along a single dimension. Confidence must be based on something and there are reasons to think the bases are eroding. Consider for instance that people no longer accept military service as an obligation of citizenship, but view military service as just another job, to be performed by volunteers. As a result, close contact with the military is no longer a common experience among the American people. One might say that a close connection between citizens and the military was an artifact of the era of the world wars, and with the passing of that era a close connection is no longer needed. Perhaps that is true. Yet it fails to notice that this is the first time—in Western Europe as well as North America—that mass democracies and professional standing armies have had to live together and depend on one another. One need not belabor the point that military culture and the democratic culture of advanced industrial societies are not readily compatible. Military leaders in the United States have expressed openly their concern that the moral values of young recruits are too low for them to be effective soldiers and sailors; and they have begun programs of moral education to repair the damage done to recruits by (what they say is) the moral laxity of civilian society.[40] The public, for its part, has been concerned (sometimes shocked) by the military's failure to embrace the values of liberal society. This concern has shown itself in controversies over gender integration and homosexual rights, in fears that the military harbors a militant and racist right–wing enclave, and in complaints that the military tries too little to protect the environment and civilian well–being when engaged in training missions. The friction between the public and the military is palpable. Yet there is little hope for reducing it while the connection between the citizenry and the military remains so loose and fragmentary.

These difficulties affect core questions about when and how the military is to deploy and, ultimately, whether and how it is supposed to fight. Distrustful of the public, yet sure that they require public support for any mission abroad, large or small, military and political leaders have become reluctant to deploy armed forces unless the mission appears certain to be a success, quickly achieved at little cost to life. Distrustful of the military, in situations where national survival is not at risk, the public has been wary of deploying armed forces abroad, unless it is for missions of humanitarian relief or "clean" police actions that prevent war from breaking out. The closer to war the deployment is, the harder the deployment is to legitimate. The public is often skeptical about what "really" justifies the military's mission and its claims for resources from society. And the military understandably believe they are risking their lives to protect the freedom of those who are personally unwilling to endure any sacrifice in their defense. Nowhere perhaps is the irony of the current situation better displayed than in Israel, where the ultra–Orthodox are most adamant in demanding a policy of militancy against the Arabs but are themselves exempt from military service in order to study Torah.[41] The difficulties armed forces face in garnering legitimacy to meet security threats are usually not so starkly posed. But neither are they uncommon. Though leading a superpower, recent presidents of the United States have been reluctant to engage in military action without securing both domestic support through Congress and international support through the United Nations, NATO, or other ad hoc venues. This was obviously true of preparations for the Persian Gulf War. It was no less true of preparations to intervene in Haiti. It is further evidence for the argument that, in new times, the sovereignty of states has declined and that that decline affects the prospects for war and peace.

A World at Peace?

When the Cold War initially ended, hope was widespread that the United Nations might at last become the focal institution it was originally intended to be, hauling down the "wild flag" of passionate nationalism, organizing efforts to enhance collective security, and containing, where it could not prevent, resort to collective violence.[42] Now, as after the end of the Second World War, that hope has dimmed. Plans for a United Nations military force, which seemed plausible in the early 1990s, faltered on the realization that the United Nations lacked the administrative capacity to conduct military operations. It also lacked the political and financial clout to raise and maintain its own forces. Major powers were willing to participate in United Nations operations and sought the blessings of United Nations resolutions to help mobilize support for their own collective security efforts. But they were unwilling to turn over command of their armed forces to officers who were outside their own chain of command. Most damaging was the perception that United Nations peacekeeping operations were not effective. That perception was based especially upon difficulties encountered in Somalia and Bosnia where belligerent forces, undeterred by the presence of U.N. peacekeeping forces, persisted in bloody violence, leaving the impression (not wholly wrong) that the United Nations was unable to impose or maintain peace.

Nevertheless, it would be unfair to conclude that peacekeeping efforts since the end of the Cold War have failed to bring peace to any region. That conclusion would ignore cases, as in Cambodia, where peacekeeping efforts were notably more successful than in Bosnia or Somalia. It also overlooks that the peacekeeping missions in Bosnia and Somalia did help reduce the amount of bloodshed and increased the flow of resources needed for humanitarian relief. To expect peace-keeping efforts to bring peace to a region may impose a wrong standard. The most serious threats to peace, we have seen, do not arise from interstate conflicts. They arise from violent reactions within weak or failed states. The problem is how to strengthen such states. External interventions by armed forces may be required. But they are difficult to organize, pay for, and justify. And they do not guarantee that strong states will be built. We do not know how to build strong states and a just peace the way we know how to build cars or computers, to do brain surgery, or conduct rocket science. We may never know.

Nevertheless, a variety of multinational institutions—not limited to the United Nations—work collectively to help contain or limit outbreaks of violent conflict.[43] Not least important are those formed to create and strengthen international laws concerning weapons of mass destruction. Over sixteen treaties have been negotiated and ratified (or observed) since 1963 that aim to limit the possible use of nuclear, biological, or chemical weapons of mass destruction.[44] One may doubt the value of legally binding treaty agreements. Treaties are often abrogated by states when convenient to pursue a particular foreign policy advantage. In any case, these treaties are difficult to enforce as there are real limits on the capacity of international organizations to monitor compliance.[45] Worse, there is the chance that nonstate terrorist groups may possess these weapons and they would be unlikely to be influenced by the niceties of law. Still the difficulties must not be exaggerated. States are not powerless, especially when acting in concert, and it is encouraging to find widespread evidence of interstate consensus about the need to limit the spread and to inhibit possible use of weapons of mass destruction.

Another important effort, championed by the International Red Cross, is to merge human rights law and the law of war into a single "humanitarian law."[46] The aim is to strengthen international norms to aid people whose government would commit atrocities against them, as has happened too often in Latin America, Africa, and Asia. Of course, asserting law is not enough. Real protection of human rights often requires armed intervention. Such interventions abridge sovereignty and may in fact be thinly veiled acts of aggression. There is no easy solution to this problem. The logic of human rights and the logic of sovereignty are not strictly compatible. Despite the difficulties, armed interventions are sometimes permissible to stop large–scale atrocities.[47] And such efforts can sometimes succeed. NATO intervention against Serbs in

Bosnia, for example, set the stage for an uneasy truce in the region. If it has yet to bring a permanent peace, it has ended the violence. The difficulty is to know more about the conditions under which these efforts are likely to succeed or fail.

That knowledge will only come from trial and error, experience and reflection. David Segal and Robert J. Waldman have shown what might be done. Their chapter in this volume reviews the history of United Nations peacekeeping operations and offers an empirical analysis to help explain the conditions under which peacekeeping operations are likely to contain the outbreak of collective violence.[48] Unsurprisingly, they find that a central element to mission success is the degree to which belligerents accept the legitimacy of the peacekeeping forces and their mandate for being there. But they also found, surprisingly, that even when legitimacy was low, peacekeepers in the area could still intervene opportunistically to gain compliance with peacekeeping agreements and limit the likelihood of renewed violence, if they were willing to use force to do so. Their important insight underlies a distinction that has been recently drawn between "traditional" and "strategic" peacekeeping missions. The first aims merely to monitor compliance with an established peace accord. It may require little more than passive observation. In contrast, "strategic" peacekeeping missions intervene where the rule of law has been abandoned and seek to impose a minimum law to protect human rights. They use force impartially against anyone who would break that law.[49] Elsewhere, I have argued that strategic peacekeeping missions should be undertaken only when there has been a clear violation of human rights law, the states proposing to intervene have the necessary resources, knowledge, and public support to back their intervention, and the odds are favorable that the intended intervention will increase compliance with human rights law.[50]

Robert L. Holmes, however, is doubtful that armed forces are ever an effective means, in the end, for keeping peace.[51] While the armed forces are used increasingly to deal with intrastate conflicts and humanitarian operations other than war, the means armed forces employ are ultimately violent. The question is whether violent means are the best available to establish national and international security. He challenges us to consider whether following principles of nonviolence may not be at least as effective as armed forces have been in establishing the just and decent society that we want to secure. Precisely because now is a time when the use of force has been thrown into doubt, we might try harder to build nonviolent institutions for conflict resolution. One might suppose that Holmes's case is unlikely to persuade many in departments and ministries of defense around the world, that it is most likely to persuade a relatively small number of academics and peace activists. Perhaps so, but only if we take a narrow view. The central assumption of Holmes's argument is that peace comes when people trust nonviolent institutions to resolve their conflicts. It is not a proposition wildly different from the one that says peace is most likely to obtain when states are strong and democratic. No hypothesis has been better proved over the last ten years than the one arguing in favor of a "democratic peace."[52]

It is by now a well-known observation that democracies tend not to wage war against one another. This is essentially an argument about how states relate with one another. Democratic states are no less likely than others to war with states that are not democratic. (That is why political leaders, when preparing the American public for war against Iraq for instance, use rhetoric that emphasizes the authoritarian character of Saddam Hussein's regime.) But when two democratic states are embroiled in a conflict, they are more likely than others to reach a peaceful settlement, and the more democratic they are, the more likely they are to settle at an early (less serious) phase of the conflict.[53] The spread of democracy helps contain the prospect of interstate war.

No less important is that democratic states may curb the rise of intrastate violence. By definition, a stable democratic state is one where internal conflicts are resolved through law, without resort to violence. Such states are less prone than others to intrastate violence, despite their higher vulnerability to terrorist attacks. They are more likely to be strong. The reason why is that they are legitimate: those who rule are accountable to those who govern and there is some sense of an overarching community under law that overrides other differences (ethnic, religious, language, etc.). When either of these conditions is absent, the legitimacy of the state is in doubt and the prospects for intrastate war are raised. That is true even for democratic states. Without legitimacy, stable democracy is difficult to maintain. This emphasis is important. As Kalevi Holsti has argued, democratic forms alone are not sufficient to curb collective violence.[54] Intrastate violence may be exacerbated by free elections in states where democratic traditions are weak and political parties are organized along ethnic or religious lines. Democratic forms must be joined with legitimacy to establish strong democracies. And as more states become strong democracies, the prospects for peace brighten. But the path toward strong democracies is long and arduous. No one knows how many miles we must go before reaching the end, assuming there is one.

Until we get there, our new times retain a connection with the past. These are not yet the days when swords are beaten into plowshares and lions lie down with the lambs. Armed forces must guard against the spread and use of weapons of mass destruction and moderate internal conflicts that plague weak or fallen states. How to organize and pay for these forces is an ongoing problem. Although the current security challenge spreads across the conflict spectrum, the use of armed forces is harder to justify. We lack a compelling narrative that links armed forces with the fate of the people and the state. The purpose for armed forces is clear: peacekeeping and deterrence. But in a time when interstate war is unlikely, for whom are these purposes being sought? This state or that? Or should we say they are sought not for the good of any one but for the good of many, and ultimately—however, naive it may sound—for the good of all? Perhaps the latter. In some sense, it is true. None are better off living under the threat of mass destruction and or in a world indifferent to the violent violations of human rights. Agreed, it is hard to mobilize resources and support for such a remote and difficult enterprise, no matter how urgent. Yet, if we recognize the hazard of letting violence run unopposed throughout the world, then we bear some responsibility to consider how it might be done. The chapters offered in this volume are an exercise of that responsibility.

Notes

1. Samuel Huntington, *The Third Wave* (Norman: University of Oklahoma Press, 1993). Whether Russia's "transition to democracy" will succeed, however, is problematic at the moment, with many observers noting that the state is weak and many institutions have fallen under the control of the Mafia. See Kalevi J. Holsti, *The State, War, and the State of War* (Cambridge: Cambridge University Press, 1996), 118, 121.
2. See Richard K. Betts, "The New Threat of Mass Destruction," *Foreign Affairs* 77 (January/February 1998): 26–41, Kathleen C. Bailey, *Doomsday Weapons in the Hands of Many* (Urbana–Champaign: University of Illinois Press, 1991).
3. Contrast the very different assumptions used to evaluate the wisdom of NATO expansion by Ronald Steel "Instead of NATO," *New York Review of Books* (January 15, 1998): 21–24 and Václav Havel, "The Charms of NATO," *New York Review of Books* (January 15, 1998): 24. More generally, see Richard l. Kugler, *Enlarging NATO: The Russia Factor* (Santa Monica, Cal.: RAND, 1996).
4. Asian reservations on this score were frankly and publicly aired at the ministerial meeting of the Association of Southeast Asian Nations (ASEAN) held during the summer of 1997. See the report of Keith B. Richburg, "Asian Leaders' Call for Human Rights Review Upsets West," *Houston Chronicle* (30 July 1997):

11A. For a sophisticated analysis of the limits to the spread of Western ideas of civil society see Ernest Gellner, *Conditions of Liberty* (New York: Penguin, 1994).

5. James J. Wirtz, "Review of The Military in New Times," *Studies in Conflict and Terrorism* 18 (January/March 1995): 72–74 and Robert E. Harkavy, "The Military After the Cold War," *Contemporary Sociology* 23 (September 1994): 638–641.

6. Identifying these trends is the main purpose of my essay, "Thinking Through the End of the Cold War," chapter 1 in this volume.

7. On this problem generally see Isaiah Berlin, *The Sense of Reality* (New York: Farrar, Straus, and Giroux, 1996), chap. 1 and Edward Shils, *Tradition* (Chicago: University of Chicago Press, 1981).

8. Several American leaders in power when the Cold War ended have remembered the discomfort they felt knowing that they could no longer rely on the old familiar assumptions about how to act and to think about acting in the international security arena.

9. Problems "managing" proxies afflicted both sides, sometimes with disastrous consequences. For the evidence see John Lewis Gaddis, *Now We Know: Rethinking Cold War History* (Oxford: Oxford University Press, 1997).

10. See chapter 2 in this volume.

11. One may point to highly visible acts, like Ted Turner's billion-dollar gift to the United Nations or to the grants awarded by George Soros, through the Soros Foundation, to promote democratization in Eastern Europe. But we should not overlook the cumulative effects of individuals acting collectively, each on a small scale, to alter world politics, whether by taking flight to become political refugees or by sending money to support Amnesty International, the Irish Republican Army, or Israel.

12. The classic statement of realist theory is Hans Morgenthau's *Politics Among Nations* (New York: Alfred A. Knopf, 1948). For reviews of realist theory see Edward A. Purcell, *The Crisis of Democratic Theory* (Lexington: University Press of Kentucky, 1973); Stanley Hoffman, "An American Social Science: International Relations," *Daedalus* 106 (Summer 1977): 41–60; and Robert L. Holmes, *On War and Morality* (Princeton, N.J.: Princeton University Press, 1989). Contemporary work relying on the realist position includes Samuel P. Huntington, *The Clash of Civilizations* (New York: Simon and Schuster, 1996); George Modelski, *Long Cycles in World Politics* (London: Macmillan, 1987); and R. Harrison Wagner, "Peace, War, and the Balance of Power," *American Political Science Review* 99 (September 1994): 593–607.

13. Alexander Wendt, "Collective Identity Formation and the International State," *American Political Science Review* 88 (June 1994): 384–396, quoted at 386. See also Bruce Russett, *Grasping the Democratic Peace* (Princeton, N.J.: Princeton University Press, 1993).

14. Holsti, *The State, War, and the State of War.*

15. In the United States, a panel of experts studying defense issues for the Congress recommends that the military focus less on fighting major wars and "pay more attention to emerging threats such as hit-and-run biological attacks on American cities." Quotation from Associated Press, "Defense Panel Recommends Paying Attention to Threats to American Cities," *New York Times* (November 29, 1997), electronic edition. In Western Europe as well, security threats are increasingly defined in terms of terrorism, drugs, fundamentalist religious or nationalist movements, and the consequences of international migration. These are not military threats as traditionally conceived. See Bernard Boëne, "A Tribe Among Tribes: Post–Modern Militaries and Civil–Military Relations," unpub-

lished paper presented at interim meeting of the Research Committee on Armed Forces & Conflict Resolution of the International Sociological Association, Modena, Italy, January 20–22, 1997.

16. Barry M. Blechman and Paul N. Nagy, *US Military Strategy in the 21st Century* (Arlington, Va.: IRIS Independent Research, 1997), 14. Robert Grant, *Counterproliferation and International Security* (Arlington, Va.: Center for Research and Education on Strategy and Technology, 1995).

17. Tim Weiner, "US Spy Agencies Find Scant Peril on Horizon," *New York Times* (January 29, 1998), electronic edition.

18. See chapter 3 below.

19. More is required than simply keeping up one's current force structures and equipment. Maintaining the current balance of power requires investments in new weapons systems based on swiftly changing electronic technologies and reconfigured forces trained to use them. A speculative account of what might be in the offing can be found in Alvin Toffler and Heidi Toffler, *War and Anti–War: Survival at the Dawn of the 21st Century* (Boston: Little, Brown, 1993).

20. Ibid., chapter 2; see also Martin van Creveld, *The Transformation of War* (New York: The Free Press, 1991) and John Keegan, *A History of Warfare* (New York: Alfred A. Knopf, 1993).

21. Holsti, *The State, War, and the State of War*, 22.

22. So Stephen A Cheney, inspector general of the Marine Corps, recently observed: "The United States has no copyright on technologies and strategies, which are easily subject to diffusion and proliferation. The next "bad guy" will be smarter— he will not play to the United States' strengths and fight it tank to tank, plane to plane. He will study its tactics and find its Achilles' heel." This comes from Cheney's article, "The General's Folly," *Foreign Affairs* 77 (January/February 1998), 157. See more generally James Burk, "Collective Violence and World Peace," *Futures Research Quarterly* 12 (Spring 1996): 41–55.

23. Barry Buzan, *People, States & Fear*, 2nd ed. (Boulder, Colo.: Lynne Rienner, 1991), chap. 2; Holsti, *The State, War and the State of War*, chap. 5; and Edward Shils, *Center and Periphery* (Chicago: University of Chicago Press, 1975), chap. 4.

24. Blechman and Nagy, *US Military Strategy*, 74–77.

25. My assumption is that modern states are likely to war with one another only when they believe they have something to gain from it. The warrants for this assumption can be found in Geoffrey Blainey's The *Causes of War* (New York: The Free Press, 1988). I do not deny that leaders often miscalculate their chances for success.

26. See chapter 4 below.

27. Other nations vary of course in their sensitivity to these issues, partly owing to their historical experience as colonial powers and partly owing to different political sensitivities, especially with respect to the issue of casualties. See Boëne, "A Tribe Among Tribes."

28. Weiner, "US Spy Agencies Find Scant Peril on Horizon."

29. Based on budget and personnel data in Department of Defense, *Annual Report–1997* (Washington, D.C.: Government Printing Office, 1997), appendices B and C.

30. These data are from Karl W. Haltiner, "The Definite End of the Mass Army in Western Europe?" unpublished paper presented at the biennial meetings of the Inter–University Seminar on Armed Forces & Society, Baltimore, October 24–26, 1997.

31. See chapter 5 in this volume.

32. This is especially true for countries in the European Union as they move to create a common currency before the end of the century.

33. Nicole E. Jaeger elaborates on this point with respect to a recent Supreme Court decision, *Loving v. United States* 116 S.Ct. 1737 (1996). In that decision, she argues, the court abandoned its usual deference before the military justice system and held that a soldier has constitutional rights that must be observed when he is court–martialed in a military capital case. See Jaeger's "Maybe Soldiers Have Rights After All!" *Journal of Criminal Law and Criminology* 87 (1997): 895–931. The courts have also been active to limit the discretion of military services to discharge soldiers or sailors whom they suspect are homosexual.

34. Joseph L. Soeters, "Value Orientations in Military Academies: A Thirteen Country Study," *Armed Forces and Society* 24 (Fall 1997): 7–32. In this article, Soeters notes the development of a multinational rapid reaction force, consisting of units from Britain, Belgium and Germany, expansion of the French–German brigade to include Spanish and Belgian forces, and the formation of the 1st German–Dutch Army Corps. Other examples could be cited.

35. *The Professional Soldier* (New York: The Free Press, 1960).

36. Eliot A. Cohen, "Civil–Military Relations," *Orbis* 41 (Spring 1997): 177–186; quotation at 178.

37. The literature on this subject has grown to be quite large. Important early articles framing the debate are Richard H. Kohn, "Out of Control: The Crisis in Civil–Military Relations," *The National Interest* 35 (Spring 1994): 3–17 and Russell F. Weigely, "The American Military and the Principle of Civilian Control from McCellan to Powell," *Journal of Military History* 57 (October 1993): 27–58. For a comprehensive review of the issues see Don M. Snider and Miranda A Carlton–Crew, eds., *U.S. Civil–Military Relations in Crisis or Transition?* (Washington, D.C.: Center for Strategic and International Studies, 1995). To consider how these events may revise our theories of civil–military relations, see "A Symposium on Civil–Military Relations," *Armed Forces & Society* 24 (Spring 1998): 375–462.

38. Ole R. Holsti, "A Widening Gap Between the Military and Civilian Society? Some Evidence, 1976–1996," working paper number 13, Project on US Post–Cold War Civil–Military Relations, John M. Olin Institute for Strategic Studies, Harvard University, October 1997.

39. In chapter 6 in this volume.

40. See Thomas E. Ricks, *Making the Corps* (New York: Charles Scribner's Sons, 1997).

41. The Israeli case is complex and defies summary in a single sentence. For an excellent study of the increasing tensions between religious beliefs and military service in Israel, see Stuart A. Cohen, *The Scroll or the Sword?* (Amsterdam: Harwood Academic Publishers, 1997).

42. One may recapture the high mood of hope that surrounded the founding of the United Nations by reading E. B. White's *The Wild Flag* (Boston: Houghton Mifflin, 1946), a collection of editorials on the subject originally written between 1943 and 1946 for *The New Yorker*. Readers of the *Letters of E. B. White*, Dorothy Lobrano Guth, ed. (New York: Harper & Row, 1989) will know that White became more skeptical as experience with the Cold War unfolded, and he was reluctant to permit reprinting or political use of the book.

43. For a description of collective peacekeeping undertaken solely by African nations, see 'Funmi Olonisakin, "African 'Homemade' Peacekeeping Initiatives," *Armed Forces and Society* 23 (Spring 1997): 349–372.

44. NATO, *Handbook* (Brussels: Office of Information and Press, 1995), 277–281 and W. Michael Reisman and Chris T. Antoniou, eds., *The Laws of War* (New York: Vintage, 1994), 66–69.

45. Current difficulties with Iraq should be sufficient to illustrate the point. For a broader discussion of the problems of controlling nuclear arms, see David Alan Rosenberg, "Nuclear War Planning," *The Laws of War*, Michael Howard, George J. Andreopoulos, and Mark R. Shulman, eds. (New Haven, Conn.: Yale University Press, 1994), 164 and Stephen J. Cimbala, "Proliferation and Peace: An Agnostic View," *Armed Forces and Society* 22 (Winter 1996), 211–233.

46. See Reisman and Antoniou, *The Laws of War*, xxi–xxii.

47. Stanley Hoffman, *Duties Beyond Borders* (Syracuse, N.Y.: Syracuse University Press, 1981), 61–73.

48. See chapter 7 in this volume.

49. Christopher Dandeker and James Gow, "The Future of Peace Support Operations: Strategic Peacekeeping and Success," *Armed Forces and Society* 23 (Spring 1997): 327–347.

50. These three conditions taken together would justify NATO intervention in Bosnia, but preclude say a NATO intervention in China to overturn suspected human rights violations there. Weighed against the costs, the merit of intervening in China, as measured by the outcomes for human rights compliance, would be impossible to calculate. NATO powers lack the resources, knowledge, or support to carry out such an intervention. This does not justify the violation of human rights of course. James Burk, "What (If Anything) Justifies Strategic Peacekeeping?" USIA sponsored lecture at the Italian Naval Academy, Livorno, October 9, 1996.

51. See chapter 8 in this volume.

52. See Bruce Russett, *Grasping the Democratic Peace: Principles for a Post–Cold War World* (Princeton, N.J.: Princeton University Press, 1993) and Michael E. Brown, Sean M. Lynn–Jones, and Steven E. Miller, eds., *Debating the Democratic Peace* (Cambridge, Mass.: MIT Press, 1996).

53. William J. Dixon, "Democracy and the Peaceful Settlement of International Conflict," *American Political Science Review* 88 (March 1994): 14–32.

54. Holsti, *The State, War, and the State of War.*

1

Thinking Through the End of the Cold War

James Burk

The end of the Cold War was a welcome event, closing successfully, from the American point of view, four decades of a taxing and dangerous military confrontation between East and West. More than once, this conflict tempted the world to resume global war and perhaps to initiate a large–scale nuclear war. That possibility is, for now, happily remote. Yet the end of this struggle has not brought unqualified relief even to those on the "winning" side. It has confused rather than clarified international relations, and it has called into question the operating assumptions which formerly justified strategic planning and public support for the armed forces. It has posed, in short, a serious challenge. The habits of thought which guided two generations of political and military leaders, of scholars and citizens alike, need to be reformed. What role will armed force and armed forces play in a world which is still turbulent, but is no longer organized by the bipolar conflict of East and West? The question may be subdivided to identify three issues of particular importance. First, what is the present character of military threats, given the shift from a bipolar to a multicentric world, and to what degree (and in what ways) does the current threat situation represent a radical break from the past? Second, how will the absence of a focal "enemy" affect military organization and, not less important, how will it affect the professional self–conception and the political and social standing of military elites who lead the armed forces of advanced industrial states? And, finally, how will a multicentric world order affect participation by national armed forces in multinational peacekeeping organizations?

Answering these questions requires sober, but speculative, analysis. The end of the Cold War and subsequent collapse of the Soviet Union has certainly changed the international order, but it has done so in a negative rather than a positive way. Sudden, unexpected, and unplanned for, the end of the Cold War has produced no social or political settlement, no concordat, to organize international relations and military operations for the foreseeable future. That does not mean that the "new world" has no order, that we are confronted by a radical break with the past. On the contrary, the central trends defining the current international, and national, situation are not completely novel, but are the result of long-term changes in technology, attitudes toward authority, and transnational organization. Operating together, these changes have affected each of the three issues we are to address. They have, first of all, multiplied the threats to which the military might have to respond while at the same time limiting the usefulness of armed force as an instrument of national policy. They have also made the successful exercise of authority more difficult, without eliminating the need for leadership, which has increased demands on military professionalism. And, they have encouraged the expansion of multinational peacekeeping forces while discouraging the use of armed force outside the framework of international coalitions, limiting the traditions of state sovereignty, and mounting pressure for nonviolent resolution of interstate and intrastate conflicts.

My aim in this chapter is to elaborate on these themes as concretely as possible, to deepen our understanding of the sociocultural forces to which the military establishment has to adapt and to indicate, in broad strokes, what the possibilities for adaptation are. To do this, we will be wise to begin by clarifying what we mean by the otherwise murky notion of a "multicentric world."

The Social and Political Context: A Multicentric World

There is nothing new in the idea that the world is divided into various centers of political and military power. Yet the idea is consequential. Those states which occupy central positions exercise vast authority, and they are jealous to maintain their high standing. States which live on the periphery, in contrast, may resent their distance from central authority if only because it seems to diminish their status and to do so, in the eyes of their leaders, without warrant. The social and ecological scarcity, which allows only a few to be at the center, a realm of plenty, and relegates the rest to relatively impoverished outer reaches, is often thought to be a major source of conflict in world affairs.[1] How conflict is organized, whether it will be violent, and what its outcomes are, these are matters of theoretical dispute. It is widely accepted, however, that nation–states, whether central or peripheral, are the major actors in the international arena. In Anthony Giddens's phrase, they are the "power containers" of the modern age.[2] They have monopolized control over political and military power. Their pursuits and their conflicts determine whatever degree of order international relations are able to assume.

Applied to international relations, the term "multicentric world" evidently refers to this untidy, but familiar structure of relative anarchy among a set of sovereign nation–states. The modern state system, we might say, fluctuates noticeably and, some would contend, systematically along a continuum that moves sometimes toward greater anarchy and sometimes toward greater order. Over the last forty years, following this logic, the state system sustained a fairly high degree of order. International affairs were organized around the bipolar structure of conflict between the United States and the Soviet Union, the only two countries during the period with legitimate claims to be superpowers. The collapse of the Soviet Union since 1989, however, has ended this regime of bipolar tension and seemed to promote a new movement toward decentralizing state powers. Rather than a bipolar international order, there is rather a proliferation of potential "poles" around which to organize the struggles among states for dominance and influence. We are moving, it seems, toward an increasingly multicentric world. Neither talk about a new world order nor the long–term development of a variety of organizations and treaty arrangements for dealing with global issues has prevented the proliferation of more autonomous regional powers—east and west, north and south—than was the case during the Cold War.[3] The difficulties posed for military planners by a world with multiple centers of power cannot be underestimated.[4] The absence of a clear "focal enemy" complicates the task of military planning, and, in democratic societies, it also makes the definition (and justification) of military missions a subject for extended and heated public debate.

Significant as these issues are, however, to focus only on recent strategic developments of the state system remains blind to more subtle and potentially more important changes in the global political order.

For James N. Rosenau, who coined the term, the idea of a multicentric world encompasses more than the strategic reorganization of interstate relations caused by the end of the Cold War.[5] It points rather to the development of a fundamental bifurcation of international politics. There is, he agrees, a realm of states as we just described it, in which states are bound by the tradition of sovereignty to pursue the interests and defend the well–being of their particular territories. But there is also a new realm of sovereignty–free actors. This realm is inhabited by multinational corporations, ethnic groups, bureaucratic agencies, transnational societies, political parties, international organizations, and even subnational social movements, who are not bound by the traditional concerns of states, who have sufficient resources to initiate global action on their own authority, and who have enough power to affect the course of global affairs. States are still important in shaping international politics, but international politics are no longer confined, as they have been, to what states do. In fact, Rosenau argues, the competence of states autonomously to manage the problems with which the international system has to deal has appreciably declined in response to three long–term trends. First, large increases in the destructive power of modern weapons systems have brought no commensurate increase in political power. On the contrary, coercion, even the threat of coercion, has become a tool of decreasing value for controlling the complex problems of domestic and international politics which states presently confront. Second, enlarging the analytic skills of individuals, providing for their higher education understood in the broadest sense, has led them to adopt a more complex and critical attitude toward the world. Compared to their grandparents, they are more questioning of traditional authority, more demanding of proof to demonstrate that authorities have in fact performed effectively, and more likely to extend their political loyalties to groups other than the nation–state. Third, technological revolutions in transportation and communication have made it feasible for sovereignty–free actors to mobilize economic, social, and political resources on a global scale. Moreover, without territories to defend or to confine them, sovereignty–free actors are not easily disciplined by any particular state. They are relatively free, much freer than was previously the case, to pursue their own goals, to give or withhold cooperation, and to shore up the sources of their own autonomy.

Not surprisingly, Rosenau sees the multicentric world as a turbulent place. It is turbulent in part because of its complexity. International politics now include a large number of actors; one has only to consider the complexity of diplomatic relations created by the increase in the number of states since the end of World War II. But Rosenau, we know, is not thinking of an increase in the number of states alone, but of the increasing number of sovereignty–free actors as well.[6] World politics are made more complex also by the increased heterogeneity of actors on the world stage, whose fates, despite their differences, are bound together by a high degree of interdependence. Moreover, this complex order is highly dynamic, and its dynamism is another source of turbulence. The dynamic quality of the world order is indicated by the frequent and rapid fluctuation of relationships among various international actors. It is caused by the high degree of variation in the goals and activities that international actors pursue.

While Rosenau has articulated these concerns more clearly than others, he is not alone in noticing the changes under foot. It bears repeating that the changes are not a direct result of the end of the Cold War. We are dealing rather with the effects of long-term trends which have been reforming the international order from the beginning of the century and long before. The persistence of the Cold War may have disguised the trends to a certain degree, permitting us to ignore their effects. But, if so, the end of the Cold War causes us to confront them head on. We have to assess the implications of a turbulent, multicentric world order for our understanding of military threats, for the organization and social standing of professional armed forces, and for the use of armed forces in maintaining a peaceful world.

The Changing Character of Military Threats

Analysis of military threats is rooted firmly in the tradition of political realism.[7] According to this tradition, world politics are dominated by states who, for purposes of international relations, may be treated as if they were unitary actors possessing a reasonably well-defined set of interests. States, furthermore, are assumed to act rationally in the pursuit of their interests. They calculate the consequences of choosing various policy alternatives and select the policy which they believe is most likely to achieve the ends they hold most dear. Of course, they are sometimes mistaken in their calculations, both about what their interests are and about how best to achieve them, but they calculate nonetheless. And they adjust their plans when they encounter (or fail to encounter) resistance from other actors in the system, modifying their aims as the cost of achieving them rises or falls. When they encounter resistance, states define it usually as a threat to their national security and so as an obstacle to be overcome. Historically, to do this, they rely on their military power and the economic foundations on which military power rests. For this reason, as Jack Levy notes, "the central proposition of realist theory is that the distribution of power in the system determines the behavior of individual states within the system."[8]

Yet if the world order is no longer state–centric as realist theory assumes, but is now (or is rapidly becoming) multicentric along the lines James Rosenau suggests, we have to ask about the continuing relevance of realist analyses of military threats. The issue is not whether realist theory is right or wrong, but how changes apparent in world reality may lead us to modify the usual ways we have of thinking about military threats and national security issues. The first question to ask is whether the end of the Cold War represents a radical break with the past, fundamentally changing national security calculations about the level and nature of military threats. As usual with important questions, this one has no simple answer.

William R. Thompson is a leading realist theorist about the causes and consequences of war. In collaboration with Karen Rasler and George Modelski, he has developed a power transition theory of structural change within the international order, a theory that focuses especially on the changing relations among global powers.[9] The value of his work for present purposes is that it adopts the long-term historical perspective which is needed to make assessments about the significance of change in the present. He begins by accepting Modelski's distinction between "global war" and "interstate war." In contrast to interstate wars, global wars are a product of structural crises which have wracked the global political system periodically over the last five centuries. They are triggered by a long period of deconcentration of control over key military capabilities as measured by control over sea (and more recently air) power. The deconcentration of military capability makes the exercise of global leadership problematic and increases competition among rising powers. A large–scale and long–lasting conflict, involving all the global powers, is the final result of this competition. In the end, when the war is done, a new leadership structure emerges with global power reconcentrated in the hands of a new state which is now in a position to establish its will (to a considerable degree) over the international order. Global wars are, in sum, wars of structural transition. They establish which state is the leading world power. And as they are relatively rare, occurring only five times since 1494, their outcomes are all the more significant.

According to this analysis, the United States emerged as the leading global power following the end of the Second World War. Since then, just as the theory would expect, the concentration of military capability in American hands has steadily declined.[10] In itself, the significance of this trend is unclear. There is no predetermined threshold of deconcentration that, once crossed, makes the outbreak of global war inevitable. Indeed, at present, the United States remains the world's leading military power. The end of the Cold War, if anything, has only strengthened the American position in this regard; the collapse of the Soviet Union meant the collapse of the only serious military challenge the United States has had to confront. Looked at in this light, the end of the Cold War does not represent a radical break with the past. The Cold War was not a war of structural transition. It surely marked the unwillingness of Soviet leaders to accept the structural outcome of World War II, which left the United States in a privileged position of power. But it did not overturn American leadership or fundamentally alter the distribution of power among states. The same states who were global powers in 1945 remain global powers today. Germany and Japan have reemerged as global economic powers following their defeat, but do not presently pose a military threat to American dominance. That said, however, the deconcentration of military power raises questions about the conditions under which global leadership is sustained. The question assumes urgency when it is remembered that the failure to retain a "unipolar" structure has five times plunged the world into global war. Interestingly, Thompson's recent work on this issue shows that a leading power's ability to maintain its share of military capability is markedly influenced by its capacity to maintain a large share of the market in leading economic sectors and high rates of growth in the development of new leading–sector technologies. It requires, in other words, that the state be able to maintain high levels of technological innovation and economic dominance.[11] Considering the mounting evidence that the United States has failed to do either one over the last twenty years, the argument may lead us to ask whether the threat of global war is increasing.

Even if analysis is confined to the realist's traditional state–centric view point, there are reasons to doubt that is the case.[12] The development of nuclear weapons distinguishes the present international system from its predecessors. Nuclear powers are much less likely than great powers of the past to go to war against one another. It has proved very difficult to translate the military superiority of nuclear power into political influence, and so the political consequences of military decline may be less than they were in the past. Although their number may decline, nuclear weapons will not disappear. They continue to serve an important defensive and deterrent purpose. Yet the destructive capacity of nuclear weapons is sufficiently great that political leaders, and the people they would have to mobilize to fight, are less likely to regard global war as inevitable. Moreover, as Donald Snow has observed, "the 'battlegrounds' of international politics...[are moving] from the trenches to the laboratories."[13] The emergence of a multicentric world, with powerful sovereignty–free economic actors, has devalued military in favor of economic power in the calculus of national security policy. The matter can be overstated, ignoring the possibility that the present situation may enhance the prospects for small interstate wars in some corners of the globe. Military threats have not all been extinguished, nor is the need for armed forces extinct. But the balance between military security and economic security has changed and the consequences of that change need to be examined.

Quite apart from the (real) dangers of nuclear confrontation, following Snow's argument, the prospects for war among the major powers are decreasing rather than increasing. The leading powers are all political democracies and allies or partners in a wide variety of projects. Under these conditions, there are no realistic scenarios which imagine conflicts among them being settled by military force. The economic expense of modern warfare is so great that advanced industrial states are reluctant to incur them. The drag on the economy is too much to bear, as the effects of waging cold war on the American and, especially, the Soviet economy seem readily to confirm. In any case, economic activities today are increasingly internationalized, so that one country seldom manufactures a product of any complexity entirely within its own borders. This is especially true for many of the goods on which modern militaries most rely to wage war, namely, high technology electronic, computing, and communication systems. One cannot easily wage war against countries on whom one depends for weapons (which is not to say that it cannot be done).

Technological innovations, Snow believes, have complicated calculations of national security in other ways.[14] Lowering costs of computing power and memory, advances in digitalizing and transmitting data, and a global movement to deregulate markets and privatize industry unleashed entrepreneurial activity and caused revolutionary transformations in economic and social relations. One important effect of these developments has been to render state borders far less important than they used to be. Business corporations are no longer necessarily bound to a particular state. They are oriented to a world economy through which they locate various parts of their operations to take advantage of geographically dispersed market opportunities. This promotes economic growth on a global scale. Yet states are confronted with the challenge of making economic actors accountable to their political community. The abilities of states to succeed in this are diminished insofar as the life chances of the international business elite are not seen to depend on their compliance with any particular state authority.[15] What power states can exert to defend their interests in this situation is typically economic rather than military. But economic power is very difficult to employ for political ends, as a review of the literature on economic sanctions will reveal, and globalization of the economy has reduced the amount of power states used to have to manipulate their own economies to serve national ends. These are disturbing trends given the dependence of state power, economic and military, on the state's ability to maintain high rates of technological growth and large market shares in the leading economic sectors.

There is no reason to assume that all states will perform equally well in this competition. On the contrary, assuming that economic development will remain uneven, it is most likely that the gap in wealth between the most and least developed regions will grow larger rather than smaller. That inequality will be less bearable in an age when instantaneous communication displays alternative ways of life, heightening the sense of relative deprivation and feelings of injustice. Even when technological progress is equally shared, the result will not necessarily be stabilizing. Increasingly sophisticated health care practices, by prolonging the useful span of life, may exert a strong generational effect reducing opportunities for mobility in third world countries whose populations, on average, are young. Given closer relations among all nations linked by the world economy, advanced states may see the growth of instabilities in the third world as a threat to their security which can be met by military force. Besides increasing the chances for small wars between major powers and third world countries, that may mean armed forces are asked to do things which they are not accustomed to doing, including peacekeeping operations, drug war interventions, civic action programs, and managing the flow of economic and political refugees.

In sum, the end of the Cold War has not brought about a radical change in the hierarchy of leading military powers established following the end of World War II. It has reduced the possibilities of a military confrontation between major powers. Nevertheless nuclear powers will retain their weapons, although perhaps in smaller numbers, and maintain the logic of deterrence. Yet radical change has been brought about by the technological revolution and globalization of the world economy. These forces have multiplied the number and importance of economic actors who are not bound by claims of national sovereignty. Their power and importance forces us to modify the traditional assumptions of political realism which have guided threat analysis. In a multicentric world, economic threats to national security increase along with economic opportunities. They have reduced the power of states to control their own economies. And, while they may have increased the probability of war with third world countries, they have reduced the usefulness of armed forces as an instrument of national power. Under these changing circumstances, the self–concept and social standing of military professionals and the organization of armed forces is bound to be transformed.

Military Professionalism and Organizational Change

The end of the Cold War has spurred a huge debate in American society about the current missions or purpose of the military establishment, about its cost, and about its proper size and organization. Claims and counterclaims in support of various positions are, not surprisingly, based on widely different beliefs about the prospects for war or the military threats which have to be met. What is happening in the United States is happening elsewhere as well, literally all around the world. If there is any general trend in the direction of change in military organizations, it is to reinforce and hasten the transition from mass armed forces, mobilized to fight particular wars, to smaller, professional armies which are continuously mobilized and possess multifaceted capabilities to respond quickly to a wide variety of threats. But it is important that we not view the debate and the prospects for change in narrow, instrumental terms, as if the only problem was the technical adjustment of military organization to the end of another war. Crucial as the end of cold war is, the debate it has sparked about the aims and organization of the military profession must be understood in light of long-term social trends. Of first importance here, is a general decline in the willingness of people uncritically to trust, accept, and follow the direction of institutional authorities. This may be part of the evolution (possibly reversible) of liberal democratic societies. But it entails a crisis of legitimacy because it makes the effective exercise of authority more difficult while it increases demands for competent leadership in a vari-

ety of spheres. Exacerbated by and characterizing the emerging multicentric world, this is an important source of social strain which has
consequences for military professionalism and organizational change.

Christopher Dandeker has documented the close historical relationship between war and military organization and the development of the
modern bureaucratic nation–state.[16] He accepts and deepens Anthony
Giddens's earlier argument that the strength of the modern state derives from its ability to monopolize and control the means of coercion
and from the development of its administrative capacities in response
to the requirements of war.[17] While sensitive to the possible abuses of a
concentrated state power, Dandeker still concludes that a global organization of strong states monopolizing the means of violence, contained
in bureaucratic military organizations, offer the most realistic basis for
achieving peace. And he is doubtful of the argument, like one James
Rosenau might make, that the power and position of nation–states has
eroded so much in recent years as to invalidate his conclusion. In an
age of global interdependence, states may be more inclined to rely on
diplomatic than military force to resolve international disputes. Nevertheless, Dandeker asserts, the state's "centralized control over the means
of violence will be a means of ensuring that whatever agreements they
make will be respected."[18] The assertion is compelling. Although it
fails seriously to consider the declining usefulness of armed force as an
instrument of policy in a multicentric world, it makes clear that there is
a continuing and important role for state authority backed by legitimate coercion. To observe the declining usefulness of military power
does not mean that military power is useless.

Nevertheless, Dandeker acknowledges that social trends current over
the last several decades make it possible to speak of the relative decline
of the military profession, not in isolation, but along with other professions.[19] The traditional social power of the modern professions is rooted,
he argues, in structural arrangements that are being eroded by major
processes of social change. There are two primary indicators of the
erosion. First is a decline of the power, which professions claim, to be
the sole provider of a service to their client. Professionals face higher
levels of competition than they have in the past. Following along with
this is an erosion of the distinctive culture or way of life which professionals were supposed to enjoy. Applied to the military case, Dandeker
points, for example, at the growing reliance by the state on civilian
defense experts and at the growing belief among military professionals
that theirs is "just another job."[20] Second is an assertion of client power

over the professional group. Clients, especially corporate clients with substantial power of their own, are less willing than they once were, simply to accept professional judgments as final judgments. Professional activity is subject to increased scrutiny and formalized social control. For the military professional this refers not only to the reluctance of public leaders to rely exclusively on military advice about the development and use of modern weapons, it includes stricter oversight and control over all resources appropriated for military use. There is, in short, a relative social devaluation of professions—to include the military profession—by a public who wish to play a more active and equal role in deciding what professional services they need and how they would like them to be delivered. What has caused this decline in professional authority?

Dandeker believes that the general decline of professionalism must be tied to the emergence of neoliberal capitalism. What he means by this term is very close to what we have been calling a multicentric world. He emphasizes the paradox that the development of a multicentric world has sustained trends toward both greater centralization and greater decentralization in the organization of economic and political life. Transnational corporations, for example, operate in a highly integrated global economy. The effect of their development, however, has been to increase central control over financial operations and strategic planning while decentralizing the production process. Politically, the multicentric world is characterized by the growth of supranational alliances and organizations and the collapse of empires, by the centralization of organized surveillance by the state over its citizens and the increased institutionalization of individual rights, guaranteeing individual autonomy and claims against central authorities. How precisely these trends are linked to the erosion of professionalism is difficult to say. But the general logic of the argument is clear. The world is increasingly organized on multiple levels, from the individual through the supranational, and in varying degrees, without a single structure able to bring coherence to the whole. We may contrast this fragmented situation of the multicentric world with the situation (somewhat idealized) of nation–states in earlier times which sustained and regulated whole, relatively self–sufficient, social organizations. In the current situation, every authority's claims for compliance to its commands and loyalty to its organization are met by competing and conflicting appeals from alternative authorities. The result is liberating to a degree, because competition among authorities supports a kind of tolerance, but it is also full of strain. Fragmented authority structures increase the difficulties of social coordination, escalating the demands for effective leadership, even as the prospects for successful leadership are curtailed.

Charles Moskos and I have clarified the implications of these developments for what we call a "postmodern" military organization.[21] The term, "postmodern," is apt. Despite much imprecision in its use in social science discourse, there is agreement that the "postmodern" condition encompasses a sense of uneasiness about the exercise of any one authority in defense of any one standard or principle, as if that was the only authority and the only standard to be defended. Ironically, it also refers to the uneasiness caused by the absence of any certain foundation to justify a set of principles, to value some as preferable to others, and to establish an authority that would work on behalf of what is preferred. It assumes the structure of a multicentric world and describes the social psychological response we have to it.

The armed forces of the developed countries moved toward the postmodern condition once it became clear that war was no longer the principal, much less the inevitable, means of resolving conflicts between them. For the American military, this movement is fairly recent and may not be accepted by all. The armed forces of Sweden, Switzerland, or Canada may best illustrate the trend. The central characteristics of postmodern military organization, *considered in pure form*, reflect the fragmentation and uneasiness of the larger society in a multicentric world. There is increased toleration of heterogeneous values, life–styles, and attitudes within and toward the military. This tolerance is shown in a variety of ways: women are more fully integrated into all military ranks and occupational specialties; more attention is paid to modify the demands of military life to accommodate the needs of the soldier's or sailor's spouse and family; homosexuals are no longer discriminated against—punished or discharged from the service—but are accepted; and conscientious objection to military service is accepted as a normal exercise of citizenship, so long as it is accompanied by performing some alternative civilian service. Military professionals no longer confine justifications of their own participation in the military to traditional military terms, but may speak in more general terms about nonmilitary forms of civic obligation. Their attitude toward arms control and disarmament is more accepting than was previously the case, in part because their perception of threats is more often defined in nonmilitary terms. Public attitudes toward the military, however, as toward all institutions of power, are more skeptical and alert than before.

This is reflected by sharp struggles over the budget, with the goal of the struggle being a reduction of defense spending. It is reflected also in the reluctance of mass media simply to accept and report what military officials say they are doing. Finally, it is reflected perhaps most significantly over the difficulty of defining or redefining the military's mission. Uncertainty over the military's mission is critical because clarity on this point lays the foundation for future planning, establishes criteria by which military performance might be judged, and provides the military with a sense of purpose. Failure to clarify the military's mission, in turn, undermines the basis for its claims to professionalism and erodes its institutional identity.

This issue of mission and institutional identity can be developed a little further. There is no dearth of missions for the military to perform. Control over and preparedness to deter nuclear war remains an important task. Some capability to fight conventional wars of varying size and with varying degrees of notice is needed as well, though there is considerable disagreement about what this entails. Deployment in support of allies is less easily justified at current levels, but will no doubt be accepted in some degree for the foreseeable future. Continuing surveillance and intelligence gathering is clearly required to verify existing and future arms control agreements, to help contain terrorist attacks, to provide early warning of troubling military operations, and so on. Sensible proposals are being defended which build on the military's capacity to engage in civic action programs, which would build on the tradition of providing infrastructure support, but operate on a global level.[22] Presumably this could include military service in times of emergency for disaster relief, whether the disaster was natural (a flood, earthquake, etc.) or man–made (a nuclear reactor meltdown, repatriation of refugees, etc.). There are, moreover, ample opportunities for peacemaking and peacekeeping operations, as we shall soon discuss. (It is worth remembering that lethal combat has become more frequent since 1945, not less, with the increase being accounted for by the growing number of civil wars.[23]) This list is not exhaustive. The point of the enumeration is to emphasize that the problem faced by the postmodern military is not the lack of things to do. Rather it is the absence of a consensus (which existed through most of the Cold War era) on some set of values or principles that provides order to the list, clearly marking some missions as focal tasks and more important to accomplish than others.

The demands imposed on military professionalism by the multicentric world are increasing rather than decreasing. That is so even if the current conditions for meeting those demands and guiding organizational change are less than propitious. The central problem is to secure a sense of institutional identity that enjoys broad public support and the support of military professionals. Solving this problem, however, will not be easy in a social setting that is increasingly skeptical of authorities of every type. If there is a way to succeed, it lies with the ability of military professionals to reflect deeply and with scholarly detachment about the needs of their own institution and its claims for social support, and then clearly and with patience to articulate for open discussion and debate the various alternatives from which the public has realistically to choose. As Moskos clearly foresaw, the military professional in the postmodern era must assume the compound and complex role of a soldier–scholar.

Prospects for Multinational Peacekeeping

Among the new missions national armed forces are taking on, one of the most important is participation in multinational peacekeeping operations. The matter has received much public attention since the end of the Cold War because its end has allowed the Security Council more effectively to exercise its powers under the United Nations charter to authorize peacekeeping operations. And opportunities to do so have not been lacking. Between 1991 and 1992, the Security Council approved new or expanded peacekeeping operations in Cambodia, what was Yugoslavia, Somalia and the Western Sahara, involving troop commitments well over 40,000. (In contrast, eight previously approved peacekeeping missions currently deploy only 11,720.)[24] In addition, the heads of state of the Security Council members met in New York on January 31, 1992 and directed the secretary general of the United Nations to recommend ways to strengthen their peacekeeping efforts and make them more efficient.[25] Their declaration acknowledged that ending the Cold War made this step possible. As usual, public rhetoric on such occasions may overstate the degree of consensus about how, or even whether, to rely more heavily on the United Nations as a vehicle for peacekeeping. In the spring of 1992, Secretary of State James Baker tried to persuade a reluctant Congress to appropriate more funds for peacekeeping efforts, arguing that support for international peacekeeping was one of the two most important initiatives in the State Department's budget (the other was aid for the former Soviet Union) and that the initiative conformed with the administration's vision of a future "characterized by a growing democratic community and the rising use of collective [i.e., multinational] engagement as an instrument of our foreign policy."[26] Yet, two days later, the leak of a draft Defense Planning Guidance for Fiscal Years 1994–1999 showed that some in the Bush administration were unwilling to embrace the strategy of collective engagement if it meant forsaking the ability of the United States to undertake unilateral military action or erosion of America's position as the leading global power.[27] The document was subsequently revised and made more compatible with Secretary Baker's position. The point remains, however, that consensus about the role of the United Nations in peacekeeping operations is far from complete.

Tensions between the competing claims of multinational organizations and sovereign states are increased, not decreased, in a multicentric world. We should not let that fact overwhelm our analysis. Nor again should we suppose that the end of the Cold War is the dominant factor at work. The basic trend in the twentieth century, which shows no signs of reversing, is to transform international warfighting coalitions into more permanent defense alliances able to keep or restore peace between nations. Early efforts at peacekeeping by the League of Nations were not notable for their success, but they were not complete failures either. There have been twenty–three United Nations peacekeeping missions since the end of the Second World War and a stronger record of success.[28] And while we have not customarily thought about it in such terms, future historians may well regard NATO as the largest and most successful peacekeeping operation of the century. Given these developments, what are the prospects for multinational peacekeeping operations to succeed? An adequate answer to the question requires both organizational and cultural analysis.

Morris Janowitz laid out the basic requirements for a peacekeeping force over thirty years ago: "It is continuously prepared to act, committed to the minimum use of force, and seeks viable international relations, rather than victory."[29] Nevertheless, peacekeeping is a relatively new role for armed forces, involving them in a special form of low intensity conflict about which relatively little is known. The aim is not to participate in the conflict, but to manage it, to keep it from becoming violent insofar as possible by relying on nonviolent means. It is not to resolve the conflict so much as to establish conditions under which the conflicts can be resolved nonviolently by others.[30] The question is how (or, for some, even whether) national armed forces can adapt themselves to perform the role effectively. Over the last decade, David R. Segal and his associates have conducted a series of studies which have shed new light on the subject.[31] While the studies focused on American Army participation in the Multinational Force and Observers in the Sinai (MFO), whose purpose was to help enforce the Camp David accords, they addressed three issues of general importance for understanding variations in the effectiveness of peacekeeping forces.

The first issue is who should participate. Early studies of peacekeeping generally accepted that multinational peacekeeping forces should only be drawn from the armed forces of middle powers rather than major powers or superpowers whose military and political entanglements might compromise the perception that they were impartial forces. But Segal established that superpower status is not a barrier to effective peacekeeping performance, which confirmed Charles Moskos's earlier finding that Britain's participation in the United Nations' peacekeeping mission in Cyprus had been effective. Acceptance of the major power deployment as neutral, however, *was* a critical precondition for effectiveness which was met in the Cyprus and Sinai cases.[32] It was not met in Beirut where American troops became targets of terrorist attack. A related question is whether combat troops should be deployed as peacekeepers, given their training as war fighters. Here again, Segal shows that combat training poses no barrier to effective role performance. Comparing the experiences of paratroopers and light infantry as peacekeepers, Segal observes that light fighters are more likely than paratroopers to believe in the value of training for peacekeeping and to believe it is a mistake to use American troops to solve other people's problems. Paratroopers maintained a more bellicose view of the world than the light fighters. But both were able to adapt to the peacekeeper role. In general, there was no relation between the strength of a soldier's combat orientation and his attitude toward peacekeeping missions.

The second issue affecting the success of peacekeeping was the problem of boredom. Peacekeeping is often a routine assignment. Ironically, the more successful the mission is, the less there is to do. This poses obvious but serious problems of morale that requires enlightened leadership to manage. The problem of boredom is mitigated to the degree that soldiers believe there is a threat to peace and to the degree that soldiers can be kept in close contact by mail or phone with their family and friends. (Yet frequent phone contact may frustrate soldiers who are unable to help family deal with urgent problems, and the calls themselves may become a source of financial strain.) Probing deeper, Segal finds that the complaint of boredom has complex and varied inner sources. The complaint is often a metaphor for the perceived loss of control over one's personal time and space, which is related to the underutilization of one's skills, lack of privacy, cultural deprivation in a foreign land, and a sense of isolation, especially in the Sinai. Measures can be taken to anticipate these problems. One might, for example, adjust the pace of training to match the pace of the jobs, allow soldiers clearly identified blocks of time that they can call their own, and overlap unit rotation to maintain the "psychosocial" history of the mission. Yet, because specific prescriptions must be tailored to fit the unique requirements of each mission, there are no simple formulas to follow. Effective peacekeeping places strong demands on innovative organizational leadership.

The last issue Segal addresses is perhaps the most interesting and difficult, namely, the cultural context for support of peacekeeping operations. While the structural strains of a multicentric world favor increased reliance on multinational peacekeeping operations, there is as yet no well–developed language to justify mobilizing people and wealth to support these endeavors. The cultural confusion has important human and organizational consequences, especially in the American case. There is a well-developed image of the American war fighter and of war fighting. There is no similar well–known stereotype that allows members of the military or the larger public easily to make sense of peacekeeping. One strains to imagine a movie about the "Blue Helmets" that would rival the "Green Berets." National cultures vary in this regard. The culture of peacekeeping is more developed in Canada than in the United States, which helps explain Canada's move to give greater weight to its participation in United Nations peacekeeping operations and less weight to its NATO involvement. But nowhere is a peacekeeping culture highly developed. So the spouses and families of soldiers and sailors are better able to understand and accept a deployment for six months on a combat–oriented training mission than they are to understand or accept a peacekeeping deployment. So are the people they know. That makes it harder for them to provide either the intimate personal or larger societal support that peacekeepers require. To a limited degree, increased participation in peacekeeping missions provides a base of experience that can help overcome this cultural deficit. The decision, in 1984, to award Purple Hearts to peacekeepers wounded in action was an important, but limited, step in this direction.

How to deal with a cultural context not yet perfectly suited to support nonviolent conflict resolution is a problem Robert L. Holmes addresses.[33] Holmes is a philosopher, and his own position that positive peacekeeping should replace war fighting is extreme. Following in the tradition of Socrates, he relies on a clearheaded mastery of the relevant tools of argument, intellectual modesty, and doubt, to provoke thought about what we take for granted. He doubts that war can be justified on moral grounds. He acknowledges that such justifications have a defensible logic. War is a response to wrongdoing, to aggression or oppression, like the Iraqi invasion of Kuwait. And war is supposed to be a means for resolving the conflict, to right the wrong or reverse the aggression. Yet the logic does not convince him.[34] Holmes distinguishes between two approaches to conflict resolution. The first approach assumes that in conflict only one side is right while the other is wrong. The matter of conflict resolution then is to assure that what is right will prevail. This is the approach that usually underlies the decision to fight wars. The second approach assumes that in conflict (even between nations) there may be right and wrong on both sides. (One can see how this may be true in the conflicts embroiling the Middle East and Eastern Europe.) Accepting this view, the matter of conflict resolution is not to see which side prevails. It is, rather, to provide conditions under which an effective solution to conflict can emerge, one that acknowledges the legitimate claims on both sides. This approach clarifies the logic of peacekeeping as it tries, in Janowitz's phrase, "to seek viable international relations, rather than victory." The question is how to institutionalize the second approach. Holmes argues that we should replace armed forces with peace forces, and that they (and we) should be trained in the arts of nonviolent conflict resolution in the same way that we have customarily trained armed forces in the arts of violent conflict resolution.

The argument is meant to be realistic and should be taken seriously, even though his recommendation, in the end, presupposes that a highly developed peacekeeping culture already exists and enjoys wide acceptance. I see no evidence of that. But, in a multicentric world there are structural forces pushing cultural developments in that direction. James Rosenau points, for example, to factors favoring development of a global culture, with worldwide norms supporting broader conceptions of self–interest than nationalism has done: nation–states cannot forego access to one another's markets and they are unable to avoid the "rippling effects" of currency crises, environmental pollution, terrorism, and so on.[35] There is, in addition, the declining utility of military power as a means for gaining political objectives or assuring national security. States are especially anxious to avoid large–scale nuclear war (or, I would add, chemical and biological war). It is clearly the case, as Holmes contends, that increasing the power of one's armed forces does not necessarily increase national security. In this context, there is a decline of what Rosenau calls "virulent patriotism," which sees the state as doing no wrong and as the unqualified object of one's political loyalties. Economic conflict, he argues, has replaced military conflict as a "battle cry" for nationalistic patriotism, but it lacks the same emotional intensity. Nor can states rely on traditional claims to sovereignty within the world community. The emergence of concern over human rights, codified in the Helsinki accords, has limited (not ended by any means) the ability of states to say they are free to do whatever they wish within their own borders. Even when violations are not censured by other states, in a multicentric world, there are other sovereignty–free actors prepared to report and condemn the matter. Pressures to put peacemaking forces in Somalia to restrain the excesses of civil war, to apply sanctions against Haiti, South Africa, and Iraq, and to deploy peacekeeping forces in Cambodia and Yugoslavia all attest to the new limits on state sovereignty. Without exaggerating the extent of these developments or their significance in a world where many other events than these are taking place, it is possible to discern both structural and normative forces that favor wider reliance on the use of multinational peacekeeping forces for nonviolent resolution of international conflicts.

Conclusions

The end of the Cold War has focused public attention on the role of armed forces in the world. The habits of thought which have guided reflection on this subject during the Cold War seem inadequate, unable to sustain a critical analysis of the present situation. The central aim of this chapter has been to introduce some new categories and themes to help identify the sociocultural forces to which the military establishment has to adapt and to indicate briefly what those adaptations might be. It is an exercise in theoretical clarification. The suggestions put forward are admittedly speculative. They speculate about the future of armed force and armed forces in an era of reform. To conclude, it will be worthwhile to restate, as succinctly as possible, the major terms and themes which, if my argument is sound, should guide our future thinking on this subject.

First, while the end of the Cold War is the catalyst for reexamination of the military's global role, we should not assume that it is somehow a positive causal force, redefining the international system in ways we have to take into account. On the contrary, the argument here is that the Cold War has at most masked basic trends which have been reforming the international system for a long time. Only now that the Cold War is ended are we having to deal directly with the consequences of these trends.

Second, total dominance of the international system by sovereign states is rapidly coming to an end. Rather than a single system revolving around the conduct of states, the global order is now multicentric. Besides sovereign states there are a myriad of sovereignty–free actors, many operating on a transnational level and many operating on a subnational level, but all able to affect the course of global affairs in ways that states cannot easily control, but to which they have to respond.

Third, in a multicentric world the utility of military power for achieving political ends is diminishing. Military power is still important. But it is less important than it was. This is partly because the destructive power of modern weapons is so great that no one can imagine any benefit coming from their use, if they were used without restriction, and it is partly because modern armed forces are so expensive that it represents a great burden on even the wealthiest societies to contemplate their use. Most important, however, is that technological revolutions in transportation and communication have so thoroughly integrated the world into one economy that the logic of interstate war is thrown into doubt. Security concerns are focused increasingly on questions of economic rather than military security.

Fourth, in a multicentric world, authority is fragmented. The claims of any one authority are often pitted against another, and the exercise of all authority is subject to greater scrutiny and challenge. This has made the successful exercise of authority by military professionals far more difficult, while increasing the demands for institutional leadership, coordination, and control. Both the institution and the profession confront an identity crisis. Everyday life in the military has become more accepting of a variety of values, life–styles, and attitudes toward its tasks. Meanwhile support for the military's mission requires endless justification and defense. It is far from true that without a cold war the military is left with no mission deserving support. Rather, in the postmodern atmosphere of a multicentric world, there is less agreement on what values should determine the relative importance of pursuing only some from the multitude of available missions and goals.

Finally, in a multicentric world, the structure of international conflicts encourages the transformation of war-fighting coalitions into permanent defensive alliances that assume the task of peacekeeping to support nonviolent conflict resolution to the fullest extent possible. Armed forces are readily adapted to perform the role. Yet care must be taken in the process. Effective peacekeeping requires that the peacekeeping force be seen as neutral by all sides to the conflict and it requires innovative leadership to offset the problems of boredom which all successful peacekeeping efforts are bound to incur. Most challenging is the problem of cultural lag associated with growth of the peacekeeping role. While world structures currently favor increased reliance on multinational peacekeeping forces, national cultures supply meager symbolic resources to enable the soldiers and sailors who act as peacekeepers, their family and friends, or even the population at large to state and justify the significance of what they are doing.

The armed forces in the current world order, in sum, are called on to exercise higher levels of military professionalism than before. The power they wield is necessary for the foreseeable future to deter any temptation to global war. Yet this power must remain in the sheath to be effective. Moreover, the states armed forces serve are increasingly besieged by their inability to manage the turbulence associated with a multicentric world. In the context of contested authority, military professionals must contribute to national and international security in economic, political, and social as much as in military terms, and do so perhaps with only minimal social and cultural support to lend meaning to the task.

Notes

1. In sociology, these ideas have been developed most completely by Edward Shils, *Center and Periphery* (Chicago: University of Chicago Press, 1975). Shils's treatment is more general and so capable of application to a greater variety of situations than the alternative "world systems" theory of Immanuel Wallerstein, *"World–Systems Analysis,"* in *Social Theory Today*, Anthony Giddens and Jonathan Turner, eds. (Stanford, Cal.: Stanford University Press, 1987). Applications of Shils's approach can be found most recently in Liah Greenfeld and Michel L. Martin, eds., *Center: Ideas and Institutions* (Chicago: University of Chicago Press, 1988).
2. Anthony Giddens, *The Nation–State and Violence* (Berkeley and Los Angeles: University of California Press, 1985).
3. Cf. Stephen D. Krasner's edited volume, *International Regimes* (Ithaca, N.Y.: Cornell University Press, 1983).
4. See John J. Mearsheimer "Back to the Future," *International Security* 15 (Summer 1990): 5–56 and Stanley Hoffmann, Robert O. Keohane, and John J. Mearsheimer, "Correspondence: Back to the Future, Part II," *International Security* 15 (Fall 1990): 191–199.
5. James N. Rosenau, *Turbulence in World Politics* (Princeton, N.J.: Princeton University Press, 1990).

6. In this century, the number of international nongovernmental organizations has increased more than tenfold, from less than 200 in 1909 to over 2,000 in 1972. See Rosenau, *Turbulence*, 409.

7. Jack Levy, "The Causes of War: A Review of Theories and Evidence," in *Behavior, Society, and Nuclear War*, vol. 1., Philip E. Tetlock, Jo L. Husbands, Robert Jervis, Paul C. Stern, and Charles Tilly, eds. (New York: Oxford University Press, 1989).

8. Ibid., 225.

9. See William R. Thompson and Karen A. Rasler, "War and Systemic Capability Reconcentration," *Journal of Conflict Resolution* 32 (June 1988): 335–366, George Modelski and William R. Thompson, *Seapower and Global Politics, 1494–1993* (Seattle: University of Washington Press, 1988), George Modelski and William R. Thompson, "Long Cycles and Global War," in *Handbook of War Studies* (Boston: Unwin Hyman, 1989), Karen A. Rasler and William R. Thompson, "War Making and State Making," *American Political Science Review* 79 (June 1985): 491–507, Karen A. Rasler and William R. Thompson, "War and Economic Growth of Major Powers," *American Journal of Political Science* 29 (June 1985): 513–538, Karen A. Rasler and William R. Thompson, *War and State Making*. (Boston: Unwin Hyman, 1989), and Karen A. Rasler and William R. Thompson, "Technological Innovation, Capability Positional Shifts, and Systemic War," *Journal of Conflict Resolution* 35 (September 1991): 412–442.

10. William R. Thompson and Karen A. Rasler, "War and Systemic Capability Reconcentration," 353.

11. Karen A. Rasler and William R. Thompson, "Technological Innovation, Capability Positional Shifts, and Systemic War."

12. Jack Levy, "The Causes of War: A Review of Theories and Evidence."

13. Donald M. Snow, *The Shape of the Future* (Armonk, N.Y.: M. E. Sharpe, 1992), 161.

14. Ibid.

15. There is no systematic survey data on the nationalist attachments of international business leaders. There is a wealth of anecdotal evidence which documents an evaluation that the regulations of national governments are decreasingly important in affecting their business strategy.

16. Christopher Dandeker, *Surveillance, Power, and Modernity* (New York: St. Martin's Press, 1990).

17. Giddens, *Nation–State and Violence*.

18. Dandeker, *Surveillance*, 226.

19. Christopher Dandeker and Paul Watts, "The Rise and Decline of the Military Profession," unpublished paper presented at the International Sociological Association, World Congress, Madrid, Spain, 1991.

20. See Cathy J. Downes, "To Be or Not to Be a Profession," *Defense Analysis* 1 (1985): 147–171 and Charles C. Moskos and Frank Wood, eds., *The Military: More than Just a Job?* (Washington, D.C.: Pergamon–Brassey's, 1988).

21. See chapter 6 below.

22. John W. De Pauw and George A. Luz, *Winning the Peace* (New York: Praeger, 1992).

23. Charles Tilly, *Coercion, Capital, and European States* (New York: Blackwell, 1990).

24. These numbers have been widely reported in the press. I take them from the *New York Times*, March 1, 1992, 4E.

25. *New York Times*, February 1, 1992, 1.

26. *New York Times*, March 6, 1992, A6.

27. *New York Times*, March 8, 1992, 1.

28. See David R. Segal and Katherine Swift Gravino, "Peacekeeping as a Military Mission," in *The Hundred Percent Challenge* (Cabin John, Md.: Seven Locks Press, 1985) and Charles C. Moskos, *Peace Soldiers* (Chicago: University of Chicago Press, 1976), 12–28.

29. Morris Janowitz, *The Professional Soldier* (Glencoe, Ill.: The Free Press, 1960), 418.

30. David R. Segal and Katherine Swift Gravino, "Peacekeeping as a Military Mission."

31. See David R. Segal, Jesse J. Harris, Joseph Rothberg, and David H. Marlowe, "Paratroopers as Peacekeepers," *Armed Forces and Society* 10 (Summer 1984): 487–506, David R. Segal and Barbara Foley Meeker, "Peacekeeping, Warfighting, and Professionalism," *Journal of Political and Military Sociology* 13 (Fall 1985): 167–181, David R. Segal, Theodore P. Furukawa, and Jerry C. Lindh, "Light Infantry as Peacekeepers in the Sinai," *Armed Forces and Society* 16 (Spring 1990): 385–403, David R. Segal, Mady Wechsler Segal, and Dana P. Eyre, "The Social Construction of Peacekeeping in America," *Sociological Forum* 7 (1992): 121–136, Larry W. Applewhite and David R. Segal, "Telephone Use by Peacekeeping Troops in the Sinai," *Armed Forces and Society* 17 (Fall 1990): 117–126, and Jesse J. Harris and David R. Segal, "Observations from the Sinai: The Boredom Factor," *Armed Forces and Society* 11 (Winter 1985): 235–248.

32. Charles C. Moskos, *Peace Soldiers*.

33. Robert L. Holmes, *On War and Morality* (Princeton, N.J.: Princeton University Press, 1989).

34. It is worth noting that the justness of the Persian Gulf War has been a subject of extended discussion and debate in philosophical circles (See the *New York Times*, February 15, 1992, 5).

35. Rosenau, *Turbulence*.

2

Armed Force and Armed Forces in a Turbulent World

James N. Rosenau

> *"I took my oath to the Soviet Union. In principle the oath is invalid. There is no Union. My country is not here."*
> —Anatoly Azovkin, 20–year-old tank division driver from the Russian Far East[1]

> *"The army is my army. For me, it's the measuring stick of my life. It's my pride, my soul's pain, my labor and sweat. It's my entire life. It's my father, who defended Leningrad, and my brothers, officers of the armed forces. What is happening to all of it?"*
> —Lt. Col. Melis Bekbasynov, member of the Russian officer corps[2]

> *"Our points of view were very different; now they have converged around peace. No one here is interested in war anymore."*
> —Padrino Pilartes, guerrilla colonel in Angola[3]

> *"A curse on you all. Where are you taking the world?"*
> —Azerbaijani grandmother shouting at a helicopter loaded with military supplies[4]

The diverse observations, cited above, reflect various perspectives from within and without the armed forces. Yet they highlight in common the enormous stresses and strains that are rocking military establishments in the aftermath of the Cold War, and they suggest common

49

questions: In a turbulent world of fragmenting polities, faltering economies, restless publics, refocused enmities, and vast international transformations where do soldiers and military organizations fit? Can they
adjust to an ever greater global complexity that is rendering the threat
and use of coercive force increasingly questionable as an instrument of
effective control? What roles and options are available to them as the
pace of change accelerates in every country and on every continent?
Must their historic mission as conservators and stabilizers of their societies give way in the face of shifting values and priorities? Can the
unique hierarchical structures and loyalties of their organizations withstand the powerful and contradictory globalizing and decentralizing
tendencies presently eroding long–standing institutions and practices
in developed and underdeveloped societies alike?

Such are the questions to which the ensuing analysis is addressed. It
rests on the underlying premise that no domain of human affairs has
remained immune to the vast transformations of our time, that it is thus
unimaginable that the huge changes at work on a global scale have not
had enormous consequences for military institutions and personnel
everywhere—that as new constraints alter the viability of armed force
as an instrument of public policy, so will changes occur in the role,
morale, coherence, and effectiveness of armed forces. No less important, it is also presumed that the consequences of turbulence experienced by militaries have, in turn, interactively fed back into the cascading
processes that sustain the vast global transformations.

Put in the rhetoric of scientific inquiry, armed forces around the world
need to be treated, not as constants in a changing world, but as both
independent and dependent variables which are central to the course of
events as the millennium draws to a close. More precisely, at any moment in time their conduct can be analyzed as both sources and consequences of global turbulence, but across time these dynamics need to
be seen as so interactive as to undermine any effort to differentiate
among causes and effects. Accordingly, here the analysis proceeds from
an assessment of armed forces as dependent variables to an estimate of
their operation as independent variables to a conclusion which probes
the interactive consequences of how their responses to change may be
shaping their emergent roles in an emergent global order.

The Turbulence Model

To facilitate these tasks a broad theoretical context is needed. Although a number of events in recent years appear illustrative of the

emergent dilemmas, opportunities, and likelihoods with which armed forces must contend in all parts of the world, the question of whether they form coherent patterns that allow for generalization on a global scale is far from self–evident. Conceivably each event is a response to unique circumstances. It is not enough, therefore, simply to call attention to the end of the Cold War, the collapse of the Soviet Union, the thirty–two nation coalition that ousted Iraq from Kuwait, or any of the many other startling developments that have rocked world politics since 1988. Episodes such as these are surely relevant to the broad changes that have ensued and that have contributed to the welter of adaptive challenges confronting armed forces. Important as they are, however, these major developments are nonetheless empirically founded and thus cannot in themselves be mined for insights as to whether and how various armed forces will cope with the adaptive challenges. The "Cold War has ended" explanation can only take us so far. It describes what no longer prevails, but it does not subsume the bases for anticipating what lies ahead.

Accordingly, insofar as the future is concerned, the end of the Cold War, the collapse of the Soviet Union, the Persian Gulf War, and other such developments must be viewed as outcomes and not as underpinnings of change. They are, in effect, the surfacing of complex dynamics which reach deep into societies and their relationships to each other. Put in still another way, outcomes are empirical and observable, whereas underpinnings are not readily discernible and can only be identified through a theoretical perspective that guides inquiry beneath the immediately visible manifestations of change. Viewed in this way, the order on which the Cold War rested did not collapse suddenly in 1989. Rather, it began its long downhill slide well before the Berlin Wall came down and the people of Eastern Europe threw off the yoke of their communist regimes. These latter developments were only the last stage in a complex process whereby the foundations of the post–World War II order underwent transformation. To be sure, pundits, politicians, academics, and people everywhere were taken by surprise when the governments in Prague, Budapest, Sofia, and other East European capitals were, suddenly, replaced. But the pervasiveness of the surprise was not so much a measure of the rapidity with which history changed course as it was a measure of how fully people tend to focus on outcomes rather than underpinnings when they respond to the course of events. Had they been sensitive to underpinnings, to the deeper sources of the events that caught their eyes, they would have appreciated well before 1989 that a new global order was in the process of evolving.

So we return to the need for a broad theoretical context which allows us to sift through the diverse empirical indicators of change and interpret them, not as responses to unique circumstances, but as products of complex dynamics that are global in scope and can thus facilitate clarity on the emergent problems with which military institutions and personnel everywhere must cope. How can we, in short, frame a theory that treats armed forces as dependent variables which, through their variation, reflect the underlying transformations that continue to rock the course of events?

In a recent book, I have developed a model that provides a positive response to this question.[5] Labeled a bifurcation, or turbulence, theory of world politics, it has the twin virtues, whatever its limitations, of being global in scope and of focusing centrally on transformative dynamics while not ignoring those that are resistant to change and promote continuity. What underpinnings have undergone fundamental transformation, the theory enables us to ask in the present context, such that armed forces around the world have experienced new challenges and evidenced new forms of behavior?

The theory's answer involves the basic parameters of world politics. If the parameters of any system are conceived as the boundaries beyond which lie the environment of the system—those recurrent patterns that may impact upon but are not a part of the system's functioning—and within which the variables of the system undergo their unceasing processes of variation, then it follows that the parameters are normally fixed features of the system. They provide its continuities in the sense that they remain constant even as fluctuations occur in its variables. They are, in effect, the foundations of global order. They are the values, premises, resources, and enduring institutions that underlie and limit the nature of the international pecking order, that accord legitimacy to issues on the global agenda, that underpin orientations toward armed force and armed forces, that justify concern for human rights, that shape predispositions toward authority and authorities, and so on through all the sources out of which variation occurs within the system.

If the parameters of world politics form the bases of the prevailing global order, then a new global order is bound to emerge if and when its parameters undergo profound transformation. And that is exactly what has happened in recent decades. For the first time since the period that culminated in the Treaty of Westphalia in 1648, the basic parameters of world politics have undergone extensive and rapid alteration, with the result that the underpinnings of a new world order have been laid.[6] Put differently, global turbulence is defined in terms of simultaneous parametric transformation,[7] a formulation which leads to the conclusion that the world is presently experiencing its first bout of turbulence in more than three hundred years.

The theory identifies three parameters that are central to any prevailing global order: the overall structure of global politics (a macro parameter), the authority structures that link macro collectivities to citizens (a macro–micro parameter), and the skills of citizens (a micro parameter).[8] Each of these parameters is judged to have undergone transformation in the current era, and the relative simultaneity of the transformations is considered a major reason why signs of an emergent global order—of deep underpinnings fostering unexpected outcomes—took politicians, journalists, academics, and others so utterly by surprise when the collapse of communism rendered them unmistakably manifest late in 1989.

Table 2.1 summarizes the changes in the three parameters, but the order of their listing should not be interpreted as implying causal sequences in which the actions of individuals are conceived to precede the behavior of collectivities. On the contrary, incisive insights into global turbulence are crucially dependent on an appreciation of the profoundly interactive nature of the three parameters—on recognizing that even as individuals shape the actions and orientations of the collectivities to which they belong, so do the goals, policies, and laws of the latter shape the actions and orientations of individuals. Out of such interaction a network of causation is fashioned that is so thoroughgoing and intermeshed as to render impossible the separation of causes from effects. Indeed, much of the rapidity of the transformations at work in world politics can be traced to the ways in which the changes in each parameter stimulate and reinforce changes in the other two.

The Micro Parameter: A Skill Revolution

The transformation of the micro parameter is to be found in the shifting capabilities of citizens everywhere. Individuals have undergone what can properly be termed a skill revolution. For a variety of reasons ranging from the advance of communications technology to the greater intricacies of life in an ever more interdependent world, people have become increasingly more competent in assessing where they fit in international affairs and how their behavior can be aggregated into sig-

TABLE 2.1
Transformation of Three Global Parameters

Parameter	From	To
Micro	Individuals less analytically skillful and cathectically competent	Individuals more analytically skillful and cathetically competent
Macro–micro	Authority structures in place as people rely on traditional and/or constitutional sources of legitimacy to comply with directives emanating from appropriate macro institutions	Authority structures in crisis as people evolve performance criteria for legitimacy and compliance with the directives issued by macro officials
Macro	Anarchic system of nation–states	Bifurcation of anarchic system into state– and multi–centric subsystems

nificant collective outcomes. Included among these newly refined skills, moreover, is an expanded capacity to focus emotion as well as to analyze the causal sequences that sustain the course of events. As will be seen, the skill revolution is also judged to have encompassed soldiers and their officers.

Put differently, it is a grievous error to assume that people are a constant in politics, that the world has rapidly changed and complexity greatly increased without consequences for the individuals who comprise the collectivities that interact on the global stage. As long as people were uninvolved in and apathetic about world affairs, it made sense to treat them as a constant parameter and to look to variabilities at the macro level for explanations of what happens in world politics. Today, however, the skill revolution has expanded the learning capacity of individuals, enriched their cognitive maps, and elaborated the scenarios with which they anticipate the future. It is no accident that the squares of the world's cities have lately been filled with large crowds demanding change.

It is tempting to affirm the impact of the skill revolution by pointing to the many restless publics that have protested authoritarian rule and clamored for more democratic forms of governance. While the worldwide thrust toward an expansion of political liberties and a diminution in the central control of economies is certainly linked to citizens and publics having greater appreciation of their circumstances and rights, there is nothing inherent in the skill revolution that leads people in more democratic directions. The change in the micro parameter is not so much one of new orientations as it is an evolution of new capacities for cogent analysis. The world's peoples are not so much converging around the same values as they are sharing a greater ability to recognize and articulate their values. Thus this parametric change is global in scope because it has enabled Islamic fundamentalists, Asian peasants, and Western sophisticates alike to serve better their respective orientations. And thus, too, the commotion in public squares has not been confined to cities in any particular region of the world. From Seoul to Prague, from Soweto to Beijing, from Paris to the West Bank, from Belgrade to Rangoon—to mention only a few of the places where collective demands have recently been voiced—the transformation of the micro parameter has been unmistakably evident.[9]

This is not to say that people everywhere are now equal in the skills they bring to bear upon world politics. Obviously, the analytically rich continue to be more skillful than the analytically poor. But while the gap between the two ends of the skill continuum may not have narrowed, the advance in the competencies of those at every point on the continuum is sufficient to contribute to a major transformation in the conduct of world affairs. More important for present purposes, the emergent global order rests on increasingly relevant micro foundations—on individuals who cannot be easily deceived and who can be readily mobilized on behalf of goals they comprehend and means they approve—a shift that tends to undermine the traditional bases of military organization and discipline. The new global order is thus more inclusive than its predecessor. Military elites retain control over resources, communications, and the other instruments through which coercive force is exerted, but increasingly they are constrained by publics and subordinates who follow their activities and are ever ready to demand appropriate performances in exchange for support.

The Macro–Micro Parameter: A Relocation of Authority

This parameter consists of the recurrent orientations, practices, and patterns through which citizens at the micro level are linked to their collectivities at the macro level. In effect, it encompasses the authority structures whereby large aggregations, be they private organizations, public bureaucracies, or military establishments, achieve and sustain the cooperation and compliance of their memberships. Historically, these authority structures have been founded on traditional criteria of legitimacy derived from constitutional and legal sources. Under these circumstances individuals were habituated to compliance with the directives issued by higher authorities. They did what they were told to do because, well, because that is what one did. As a consequence, authority structures remained in place for decades, even centuries, as people unquestioningly yielded to the dictates of governments or the leadership of any other organizations with which they were affiliated. For a variety of reasons, including the expanded analytic skills of citizens noted above, the foundations of this parameter have also undergone erosion. Throughout the world today, in both public and private settings, the sources of authority have shifted from traditional to performance criteria of legitimacy. Where the structures of authority were once in place, in other words, now they are in crisis, with the readiness of individuals to comply with governing directives being very much a function of their assessment of the performances of the authorities. The more the performance record is considered appropriate—in terms of satisfying needs, moving toward goals, and providing stability—the more are they likely to cooperate and comply. The less they approve the performance record, the more are they likely to withhold their compliance or otherwise complicate the efforts of macro authorities.

As a consequence of the pervasive authority crises, states and governments have become less effective in confronting challenges and implementing policies. They can still maintain public order through their police powers, but their ability to address substantive issues and solve substantive problems is declining as people find fault with their performances and thus question their authority, redefine the bases of their legitimacy, redirect their loyalties, and withhold their cooperation. Such a transformation is being played out dramatically today in the former Soviet Union (as it did less than two years earlier within all the countries of Eastern Europe), where the redirection of loyalties and the application of performance criteria have eroded favorable predispositions toward the military establishment. In the words of one Russian,

> In our parents' day, [the army] could get away with [a compulsory draft]. People believed in suffering and sacrifice then. They believed they could do everything for the glory of the motherland and get no compensation for it. That's not a good argument for our generation.[10]

But authority crises in the former Soviet Republics are only the more obvious instances of this newly emergent pattern. It is equally evident elsewhere, albeit the crises take different forms in different countries and different types of private organizations. In Canada, the authority crisis is rooted in linguistic, cultural, and constitutional issues as Quebec seeks to secede or otherwise redefine its relationship to the central government. In France, the devolution of authority was legally sanctioned through legislation that privatized several governmental activities and relocated authority away from Paris and toward greater jurisdiction for the provinces. In China, the provinces enjoy a wider jurisdiction by, in effect, ignoring or defying Beijing. In Yugoslavia, the crisis led to violence and civil war, until the state itself dissolved. In the crisis–ridden countries of Latin America, the challenge to traditional authority originates with insurgent movements or the drug trade. And in those parts of the world where the shift to performance criteria of legitimacy has not resulted in the relocation of authority—such as the United States, Israel, Argentina, the Philippines, and South Korea—uneasy stalemates prevail in the policymaking process as governments have proven incapable of bridging societal divisions sufficiently to undertake the decisive actions necessary to address and resolve intractable problems.

Nor is the global authority crisis confined to states and governments. They are also manifest in subnational jurisdictions, international organizations, and nongovernmental transnational entities. Indeed, in some cases the crises unfold simultaneously at different levels: just as the issue of Quebec's place in Canada became paramount, for example, so did the Mohawks in Quebec press for their own autonomy. Similarly, just as Moldavia recently rejected Moscow's authority, so did several ethnic groups within Moldavia seek to establish their own autonomy by rejecting Moldavia's authority. Similarly, to cite but a few conspicuous examples of crises in international and transnational organizations, UNESCO, the PLO, the Catholic Church, and the Mafia have all experienced decentralizing dynamics that are at least partly rooted in the replacement of traditional with performance criteria of legitimacy.

The relocating of authority precipitated by the structural crises of states and governments at the national level occurs in several directions, depending in good part on the scope of the enterprises people perceive as more receptive to their concerns and thus more capable of meeting their increased preoccupation with the adequacy of performances. In many instances this has involved "downward" relocation toward subnational groups—ethnic minorities, local governments, single–issue organizations, religious and linguistic groupings, political factions, trade unions, and the like—which, in turn, have become increasingly aware of their growing coherence and clout, an awareness that elsewhere I have suggested has brought the world into a vigorous age of subgroupism.[11] In some instances the processes of authority relocation has moved in the opposite direction toward more encompassing collectivities that transcend national boundaries. The beneficiaries of this "upward" relocation of authority range from supranational organizations like the European Community to intergovernmental organizations like the International Labor Organization, from nongovernmental organizations like the Greenpeace to professional groups such as Médecins sans Frontièrs, from multinational corporations like IBM to inchoate social movements that join together environmentalists or women in different countries, from informal international regimes like those active in different industries to formal associations of political parties like those that share conservative or socialist ideologies—to mention but a few types of larger–than–national entities that have become the focus of legitimacy sentiments. Needless to say, these multiple directions in which authority is being relocated serve to reinforce the tensions between the centralizing and decentralizing dynamics that underlie the turbulence presently at work in world politics.

In short, the emergent global system rests on an increasingly fluid pecking order. Although still hierarchical in a number of respects, the weakening of states and the pervasiveness of authority crises has rendered the pecking order more vulnerable to challenges and more susceptible to changes. Put differently, with states weakened by paralysis and stalemate, the power equation underlying the pecking order has been substantially altered. Where raw elements of power—armies, oil deposits, agricultural production, etc.—were once the major terms of the equation, now their values have declined relative to such complex terms as societal cohesion, the capacity to draft soldiers, decisiveness in policymaking, and the many other components of a country's ability to surmount authority crises and avoid paralyzing political stalemates.

It follows that any organization founded on clear lines of command and unquestioned compliance with orders is presently being undermined. Such a process is already evident in perhaps the most disciplined organization extant today, the Mafia,[12] and it can also be discerned in many, if not all, armed forces. Or at least it is sufficiently manifest to conclude, as will be seen, that the transformation of the macro–micro parameter has weakened military establishments as well as the states they serve.

The Macro Parameter: A Bifurcation of Global Structures

For more than three centuries the overall structure of world politics has been founded on an anarchic system of sovereign nation–states that did not have to answer to any higher authority and that managed their conflicts through accommodation or war. States were not the only actors on the world stage, but traditionally they were the dominant collectivities; they set the rules by which the others had to live. The resulting state–centric world evolved its own hierarchy based on the way in which military, economic, and political power was distributed. Depending on how many states had the greatest concentration of power, at different historical moments the overall system was varyingly marked by hegemonic, bipolar, or multipolar structures.

Today, however, the state–centric world is no longer predominant. Due to the skill revolution, the worldwide spread of authority crises, the impact of dynamic technologies, the globalization of national economies, and many other factors, it has undergone bifurcation.[13] A complex multicentric world of diverse, relatively autonomous actors has emerged, replete with structures, processes, and decision rules of its own. The sovereignty–free actors of the multicentric world consist of multinational corporations, ethnic minorities, subnational governments and bureaucracies, professional societies, political parties, transnational organizations, and the like. Individually, and sometimes jointly, they compete, conflict, cooperate, or otherwise interact with the sovereignty–bound actors of the state–centric world.[14] Table 2.2 delineates the main differences between the multicentric and state–centric worlds.

While the bifurcation of world politics has not pushed states to the edge of the global stage, they are, to reiterate, no longer the only key actors. Now they are faced with the new task of coping with disparate rivals from another world as well as the challenges posed by counterparts in their own world. A major outcome of this transformation of macro structures is, obviously, a further confounding of the hierarchical arrangements on which world politics is based. Not only have authority crises within states rendered the international pecking order more fluid, but the advent of bifurcation and the autonomy of actors in the multicentric world so swelled the population of entities that occupy significant roles on the world stage that their hierarchical differences were scrambled virtually beyond recognition well before the end of the Cold War intensified the struggle for international status.

TABLE 2.2
Structure and Process in the Two Worlds of World Politics

Factor	State–Centric World	Multicentric World
Number of essential actors	Fewer than 200	Hundreds of thousands
Prime dilemma of actors	Security	Autonomy
Principal goals of actors	Preservation of territorial integrity and physical security	Increase in world market shares, maintenance of integration of subsystems
Ultimate resort for realizing goals	Armed force	Withholding of cooperation or compliance
Normative priorities	Processes, especially those that preserve sovereignty and the rule of law	Outcomes, especially those that expand human rights, justice, and wealth
Modes of collaboration	Formal alliances whenever possible	Temporary coalitions
Scope of agenda	Limited	Unlimited
Rules governing interactions among actors	Diplomatic practices	Ad hoc, situational
Distribution of power among actors	Hierarchical by amount of power	Relative equality as far as initiating action is concerned
Interaction patterns among actors	Symmetrical	Asymmetrical
Locus of leadership	Great powers	Innovative actors with extensive resources
Institutionalization	Well established	Emergent
Susceptibility to change	Relatively low	Relatively high
Control over outcomes	Concentrated	Diffused
Bases of decisional structures	Formal authority, law	Various types of authority, effective leadership

Source: Reproduced from James N. Rosenau, *Turbulence in World Politics: A Theory of Change and Continuity* (Princeton, N.J.: Princeton University Press, 1990), 250

The New Global Order: Enduring or Transitional?

However the new global order may be labeled—and the designation is not a trivial exercise[15]—one conclusion derived from the foregoing analysis stands out. While the new order is marked by a high degree of uncertainty and dynamism, these features may not be transitional. It is tempting to conclude otherwise, to view the present era of turbulence as temporary, as a necessary process through which parameters get redefined and settle into a new set of premises. After all, it could well be reasoned, the ideational foundations of a new order and the habitual modes on which it rests do not automatically fall into place, so that the period between the end of one order and the evolution of a new one is bound to be disruptive as people cling to old ways of doing things in the face of new circumstances. Time is needed for experimentation, for trial and error, for sorting out alternatives, for actors with different learning curves to arrive at the same plateau on which their new relationships can flourish. Accordingly, the argument concludes, it will surely require a long transitional period before the outlines of the new order, much less the order itself, come fully into focus. Indeed, it may even take a while before it is clear that the old order has fully passed into history.

There is surely much validity in this reasoning. The attenuation of mental sets, the collapse of paradigms, and the abandonment of structural constraints certainly occur with much greater speed than the subsequent processes through which new orientations and structures are formed and become deeply rooted. History brilliantly affirms that it is easier to destroy institutions than it is to construct their replacements. Nevertheless, it may well be that today's uncertainties and dynamics are not as temporary as logic dictates and as observation suggests, that the changes at work in world politics are such that there is no reason to expect the slow evolution of practices and processes on which the stabilities of a new order will rest.

Conceivably, in other words, the uncertainties and dynamics of the present period are permanent rather than transitional features of the new order. Why? Because the parametric transformations are all in the direction of enduring commotion rather than stable patterns. They do not point to the emergence of new political entities that can supplant the state as effective means of achieving societal cohesion and progress. They do not hint at the development of global structures which can accommodate both the decentralizing dynamics operative within the state–centric world and the centralizing dynamics at work in the multi-centric world. They do not in any way suggest that more analytically skillful citizens and publics are likely to be impressed with the intractability of most of the issues on the global agenda and thus accepting of leaders whose performances are bound to fall short of their expectations. Rather the central tendency in the case of each parameter involves movement toward end points that are inherently and profoundly pervaded with uncertainty and dynamism, with impulses to sustain change rather than settle for new equilibria, with a potential for restless dissatisfaction with power balances and continuing resistance to whatever pecking order may seem to prevail.

In short, the most powerful forces underlying the emergence of a new global order are all conducive to persistent, long–term tensions between the need for more centralized international institutions and the equally compelling need to develop more decentralized domestic institutions. These tensions can be readily discerned in every part of the world today, in the tendencies toward strengthened regional organizations (e.g., in Europe and Latin America) and in those fomenting authority crises within countries. The fact that the Soviet Union and Yugoslavia came apart internally even as both sought to become increasingly linked externally to international organizations offers quintessential examples of how fragmentation and integration are both central characteristics of the new global order. And these are only the more conspicuous of a multitude of illustrations that could be cited.

Armed Forces in a Turbulent World

Although brief, the foregoing conception of the dynamics of present–day global turbulence enables us to formulate incisive questions with which to probe military organizations as dependent variables: what changes have they undergone with the advent of a bifurcated world of weakened states, intensified subgroupism, authority crises, and more analytically skillful citizens? Have their strict hierarchical foundations and their rigorous commitment to disciplined command structures enabled them to withstand the strong currents of turbulence? Or have they, too, been swept along by the powerful forces of change? Can signs of the skill revolution be discerned in the behavior of enlisted men and officers? Are military organizations undergoing crises of authority and turning to forms of subgroupism? Has the ever–expanding complexity of a bifurcated world resulted in a greater diversity of tasks undertaken by armed forces? In short, have they managed to remain essentially constant—beacons of stability in a fragmenting, unstable world—or is their conduct expressive of discernible and significant variation, thus affirming our underlying premise that no domain of human affairs has remained immune to the vast transformations of our time?

Developing initial responses to these questions perforce requires reliance on anecdotal data. A number and variety of empirical indicators have surfaced in recent years, but systematic inquiries derived from the turbulence model have yet to be undertaken. Unsystematic as they may be, however, the available data are sufficiently suggestive to allow for tentative interpretations and the framing of hypotheses designed to generate more rigorous findings.

At first glance, that is, it seems clear that change is stirring upheaval within military institutions. Notwithstanding adherence to the cannons of analytic caution, one cannot help but be impressed by the continuing stream of events indicative of how the transformation of all three parameters has introduced variability into military affairs. Indeed, they readily highlight succinct answers to the foregoing questions: No, the indicators suggest, militaries in most countries have not been able to withstand modifications of their strict hierarchical foundations and disciplined command structures. Yes, signs of the skill revolution are manifest in the behavior of military personnel. Yes, military organizations virtually everywhere are undergoing authority crises and resorting to forms of subgroupism. Yes, many armed forces are taking on a diversity of new tasks. And yes, there are good reasons to continue to presume that the military have been no more immune to the dynamics of turbulence than any other societal institution.

In order to elaborate and justify these overall interpretations, the pervasive traces of parametric transformation evident in the conduct of military personnel and the structure of military establishments are examined in the context of a few dimensions of each parameter. In the case of parametric changes at the micro level, the ensuing analysis probes how expanded analytic skills have affected the orientations of officers and enlisted personnel alike and how the need for sustaining the skill revolution has posed a delicate challenge to the command structure of military organizations. The transformation of the macro–micro parameter is explored in terms of the impact of global authority crises upon the exercise of military discipline and the capacity of armed forces to maintain and replenish their ranks. The changes stirred by the bifurcation of the macro parameter are briefly assessed in terms of old roles being shed and new tasks being shouldered by armed forces, the constraints imposed by the emergence of global norms and vigorous transnational social movements, and the tendencies toward subgroupism within and across military ranks.

Since the transformative consequences of each parameter are interactive, as previously noted, the order in which we assess the various dimensions of the military as a dependent variable matters little and should not be viewed as reflecting an assumption as to which change dynamics are primary and which are secondary. The analysis begins with transformations at the macro level because these encompass the broadest array of phenomena, but this is not to imply that the macro circumstances with which armed forces must cope are the prime determinants of their conduct.

Diminished Roles and New Tasks

The bifurcation of world politics is conceived as reflecting an ever–growing interdependence in which decentralizing and centralizing tendencies at all levels of political aggregation reinforce, offset, or otherwise sustain global affairs. One consequence of this complexity is that armed forces are becoming more marginal even as their capabilities have become more relevant. That is, the weakening of national states, the proliferation of subnational organizations and groups in the multicentric world, the shift of systemic agenda away from security and toward economic issues, the redirection of numerous adversarial relationships within regions and countries (as well as between former superpowers), the contraction of military budgets—to mention only a few of the dynamics at work—has led societies to be less reliant upon and less respectful of their military institutions even as the very complexity that undermines their reliance leads them to seek out their militaries to perform new tasks that are also a consequence of greater complexity and extensive change.

This paradoxical pattern of lesser reliance and expanded tasks derives, in effect, from the combination of economic recession, war weariness, and shifting norms which make the conduct of military operations less attractive. No longer are elites and publics inclined to view military actions as a solution to problems and, indeed, in some cases such actions are considered to be a major source of the difficulties with which people have to contend.[16] High–tech weapons may allow for the quick defeat of a menacing enemy, but they are of no value in controlling the many enduring problems that continue to unfold away from the battle-field. It is no accident, therefore, that a number of wars have ground to a halt in recent years; each of these various situations had its own unique dynamics, but it seems clear that each also ran afoul of humankind's lessened tolerance for prolonged combat and its greater appreciation of the limits of military effectiveness.[17] So there is less for armed forces to do—to protect, to threaten, to prepare for—and as a result their roles in governmental decision making and societal affairs have undergone considerable attrition. On the other hand, the very same combination of economic recession, war weariness, and shifting norms has given rise to a diversity of challenges which only the military, being well orga-nized and still in command of substantial resources, seem competent to meet. Troops of NATO, for example, were used to deliver emergency supplies to the people of the former Soviet Union. Similarly, the U.S. military was recently asked to take on the task of preparing shelters, sanitary facilities, and medical care for thousands of Haitians fleeing their country, a role that was widely viewed by defense officials as "a headache" but that was nonetheless accepted.[18] Subsequently, a similar request to take on a new leadership role in the war against drugs, made precisely because of its expertise in creating "a unified command au-thority," was rejected by the U.S. military, and one of the reasons for the rejection suggests that the adaptation of service personnel to their diminished roles and new tasks is far from easy:

> Its reluctance now to take on a bigger role was described...as a consequence, in part, of the Persian Gulf War, which made some military officers scornful of mere anti–drug operations. But it was said to reflect also a Pentagon wariness about becoming too closely identified with the failure to make inroads against a poten-tially intractable problem.[19]

In addition to new roles in the fields of humanitarian assistance and drug control, militaries in various parts of the world are increasingly likely to be called upon to take on peacekeeping responsibilities. With the dynamics of bifurcation and shifting values toward international organizations thrusting the United Nations into the limelight as an in-strument for coping with both intrastate and interstate conflicts,[20] more and more countries will be asked to contribute personnel to such mis-sions. For those involved these new supranational tasks are likely, when combined with the growing incentives to honor subnational values, to confound further the increasingly complex question of the goals and purposes for which national military organizations are designed to serve and sacrifice (see below). Not only have militaries had to adjust to an obfuscation of their enemies, but the advent of turbulence in world politics has also beclouded the identity of their superiors.[21]

Global Norms and Social Movements

The spread of global norms and the greater coherence and effectiveness of social movements constitute additional constraints on the military stemming from the transformation of the macro parameter. More precisely, one norm, human rights, and several social movements, particularly the peace, women's, and environmental movements, have tended to narrow the freedom of action that armed forces have historically enjoyed. The efforts of the Organization of American States to hold members of the Haitian military accountable for their atrocities, the finding of Mexico's National Commission on Human Rights that a senior army general "bore the major responsibility" for the deaths of narcotics agents,[22] and the court convictions of former East German border guards for explicitly violating the human rights of would–be escapees are recent examples of how a growing global concern for rights of individuals is subjecting military personnel to new forms of scrutiny and condemnation.[23] It seems probable, too, that the increasing reluctance (noted below) of military officers to order their troops to fire on fellow citizens gathered for protests in town squares is, at least in part, a consequence of the emergence of human rights as a global norm.

In a like manner, the pressures generated by various social movements, all of them transnational in scope and global in appeal, have added to the limitations within which armed forces must operate.[24] Just as the peace movement contributed to the diverse challenges to military authority discussed below, so has the worldwide preoccupation with the role of women encouraged many countries to open combat roles in their officer and enlisted ranks to females. Between the late 1960s and the early 1990s, for example, the proportion of women in NATO forces rose from 2 percent to 7.4 percent,[25] a pattern that has presumably been further reinforced by the exemplary performance of American women soldiers in the Persian Gulf War. At the same time it is also clear that the adjustment of military men to the arrival of female compatriots has not been easy. To cite one relevant instance, the feminist movement has reshaped the macro context in which military routines are sustained. The Corps of Cadets at Texas A&M University, for example, was charged with mistreatment of its female members and for the first time an outside group was given authority to examine the corps' policies and to suggest policy changes.[26] Nor is this an isolated example, as shown by the national controversy stirred by the gross treatment of women by Naval aviators at the annual convention of the Tailhook Association.[27]

Subgroupism

The opportunities afforded by the evolution of the multicentric world, with its incentives to seek subgroup coherence and privileges, have not been lost on more than a few armed services. In effect, many appear to have broken a long tradition in which they operated as a tightly knit elite acting quietly behind closed doors, pressing policies, pulling levers, stressing limits, or otherwise exercising influence. Historically, that is, the military has acted as an agent of the state, either protecting its civilian leaders from perceived adversaries or, in some instances, replacing them on behalf of perceived needs for order. With the transformation of the macro parameter, however, the military have moved from behind closed doors into the public arena, acting not as agents of the state but as claimants on its resources, much like the other subgroups that populate the multicentric world.

In short, where subgroupism within armed services used to refer to the cohesion of a governmental elite, now it also connotes the plaintive demands of a besieged subgroup outside the prevailing power structure. Where the military could readily co–opt other groups in the private sector, now they have to compete with them; and, in so doing, they have also come to emulate them. Notwithstanding the constraints of a long–standing commitment to organizational discipline, military leaders have been increasingly emboldened by the diminution of their funds, prestige, and tasks to seek redress through the rowdier methods of aggregation used by other actors in the multicentric world. In Russia, for example, officers came together to concert their demands, which included the issuance of a call for coverage by nationwide television and the formation of a "coordinating council" to represent their interests.[28] While some might dismiss the subgroupism of the military in the former Soviet Union as a special case resulting from a particularly sudden and unexpected set of events, it is equally plausible, given the bifurcation of global structures, that subgroupism among militaries elsewhere is likely to increase in proportion to their loss of status, support, and perquisites.[29]

This is not to say, of course, that turbulent conditions have brought an end to the readiness of all militaries to eschew use of their instruments of coercion in favor of conventional subgroup behavior. Both successful and failed military coups d'état will doubtless continue to mark the political scene—if only as last, desperate measures to cling to privilege. But the politicization (can we dare call it a demilitarization?) of armed forces does seem likely to grow as their roles diminish and as the war weariness of publics deepens. Viewed in this way, the recent decisions taken by the military within both the government and revolutionary movement of El Salvador to reverse course and compete through political parties and elections stands out as a dramatic indicator of this underlying global trend.

The Redefinition of Interests

Turning now to the transformation of the micro parameter, it is useful to stress again that all the parametric changes are interactive. While the orientations of individuals in the military have certainly been shaped by the foregoing macro circumstances, so has the enlargement of their analytic skills depicted below resulted from the operation of dynamics at the micro level which, in turn, have been a source of the macro changes.

Among the numerous indicators of the skill revolution having an impact on armed forces, perhaps the most noticeable consequence of a transformed micro parameter can be found in the expansion of the skills with which both enlisted and officer personnel assess their own interests. Traditionally military persons devoted little energy to pondering their personal interests. Rather, a deep-seated habit prevailed through which self–interests were equated with service and national interests. Now, however, there are numerous signs that enhanced analytic skills have led to a disaggregation of the concept of self–interest. If the concept of expanded analytic skills is operationalized as a greater capacity to play out scenarios that locate where individuals fit in the processes of world politics,[30] it is not difficult to uncover widespread indicators of members of armed forces being preoccupied with their changing roles. For the most part, this preoccupation takes the form of distress over the loss of support, tasks, and perquisites, and in turn the distress is most manifest in a seemingly increasing readiness of servicemen and women to attach higher priority to their own welfare than to the military organizations or countries they serve. The plaintive lament of the Russian colonel quoted among the opening epigraphs is echoed not only in the words of many individuals in the military of the former Soviet Union, but comparable statements can be found in the expressions of counterparts in other countries that are scaling back their armed forces in the face of budget cuts and the end of Cold War rivalries. Consider for example, the words of U.S. Vice Admiral Leighton W. Smith, Jr., the deputy chief of naval operations: "My biggest concern is they'll pull the rug out from under us and we'll go into an undisciplined free fall."[31] Or ponder the observation of Roger Spiller, a professor of military history at the U.S. Army's Command and General Staff College: asked for a reading of the present mood of the military, he responded that "I think there's a good deal of anxiety. They'd have to be completely insensitive not to feel anxiety about the impending reductions in the Army. These are men and women acutely aware of their surroundings, which right now are rather hostile to their future plans."[32]

The growing preoccupation of military personnel with their own welfare is, of course, a violation of the traditional stereotype wherein the good soldier is supposed to accept unthinkingly whatever decisions their military and civilian superiors make with respect to their status and activity. This conception of military command structures lies at the very heart of combat readiness and effectiveness. Yet, nowhere can one find traces of it being voiced by either government officials and top ranking officers. Some may be good soldiers in the sense that they are "remarkably philosophical" and "understand what is happening and the reasons for it,"[33] but few if any have publicly affirmed that the military is duty–bound to accede to the changes initiated by their political superiors. Indeed, it is a measure of how profoundly the military have been caught up in the parametric transformations at work in world politics that all concerned, including those of us who analyze military institutions, take for granted that military personnel are no longer silent and obedient with respect to their own welfare. Not long ago, for example, the effort of Russian President Yeltsin to win over disgruntled officers by publicly offering them 120,000 apartments and 70,000 quarter–acre plots[34] would have seemed absurd, a form of public bargaining entirely out of place in a military context. Today, however, such an offer seems more commonplace than anomalous, a cogent indicator of how fully the analytic skills of military personnel are no longer treated as a constant in the political equation.

The most obvious trace of the skill revolution's impact is to be found in fragmenting countries, where the ethnic identification of military personnel is fostering the same kinds of divisions among them that divides their societies. The army in Algeria, for example, is considered to be "crossed by the same currents that cross Algerian society," with the top command "being dead set against the [Muslim] fundamentalists," but this perspective changes "as you travel down the ranks."[35] Similarly, Ukraine's insistence that members of the armed forces stationed on its territory sign loyalty oaths bespeaks a recognition that enlisted personnel and officers are no longer dutiful, that they are inclined to analyze their situations, sort out their loyalties, and then to follow the course most appropriate to their personal commitments. And that is exactly what transpired. As one analyst put it, military units and many officers are "following their paychecks" to republican governments, with the result that "Soviet conscripts, more and more, are serving within the boundaries of their home republics in a process of homogenization."[36] But it is a homogenization that stems from decentralizing dynamics, the potential of which is poignantly revealed by the fact that the Black Sea Fleet of the former Soviet Union consisted of people of forty-six nationalities and that some of its ships had sailors of as many twenty-five nationalities.[37] Put even more poignantly, with the ethnic fragmentation of the Soviet military, "fathers are worrying about facing their soldier sons across a battlefield, if they end up serving different states. Officers are watching the unity of their troops give way to ethnic enmity."[38]

It would be a mistake, however, to view the disruptive consequences of the greater capacity of military personnel to extend their analytic skills beyond the boundaries imposed by the canons of strict military discipline as confined to the enlisted ranks. The unity of top military leaders has also been splintered by the skill revolution. Or at least there are good indications that high–ranking officers are no more uniform in their underlying political attitudes than any other occupational elite, that the new license enjoyed by military personnel of all ranks to think analytically—to locate for themselves where military institutions fit in the panoply of societal dynamics—would appear to be as much a source of divergent perspectives as of widely shared consensus among leaders of defense establishments. Even as many senior officers of the former Soviet Union, for example, gave expression to decentralizing tendencies and opted to give higher priority to their ethnic loyalties, so did others articulate centralizing values and pressed for the maintenance of the military as a single, united command structure. Likewise, just as some took a right–wing position favoring the use of strong–arm techniques to restore order, so did others resist the militant posture and voiced support for democratic procedures.[39]

In short, the enhancement of analytic skills among military personnel seems likely to undermine rather than reinforce traditional command structures. The roles played by top Soviet military leaders on both sides of the August coup and its subsequent failure is perhaps even more eloquent testimony against the presumption that military elites have somehow managed to avoid the divisive consequences of a transformed micro parameter.

Technology and the Expansion of Analytic Skills

Nor is the evidence of greater analytic skills on the part of military personnel due only to the way in which turbulent conditions have heightened their self–interests. The increasingly technological nature of warfare has also contributed to their capacity to play out self–interested scenarios. High–tech weaponry cannot be operated by poorly educated and unskilled soldiers or sailors. Considerable advanced training is required to man modern weapons, thus necessitating that military institutions convert their recruits from physically able individuals to skilled technicians. And once they do, once enlisted personnel are competent to use and repair computers, complex targeting devices, intricate photographic equipment, and the other modern devices through which the instruments of war deliver their loads, the raw recruit is no longer an unthinking individual incapable of pondering cause and effect or discerning the diverse routes through which desired outcomes are produced. Rather, having been tooled up in high–tech forms of warfare, he or she is possessed of new analytic capacities that can readily be extended into other, more political realms.[40]

A quintessential insight into the relevance of the skill revolution for military personnel is provided by the spate of TV advertisements in which the U.S. armed forces stress the educational opportunities of military service. Designed to recruit volunteers, the advertisements concentrate on the subsequent, postservice uses to which technical skills acquired in the armed forces can be put, the underlying message being that one leaves the services a much more analytically competent individual than one was before joining. The same insight is implicit in the complaints of former Soviet officers who were assigned young non–Russian–speaking draftees from Central Asia to operate complex, modern missile equipment.[41]

For closed and authoritarian societies like China the necessity of implementing a skill revolution poses a difficult challenge. Indeed, it has been the focus of considerable debate among Chinese leaders. On the one hand, the importance of high–tech weapons in the success of the U.S.–led coalition against Iraq made it clear to them that they would have to bring the skill revolution to their own armed forces if they wanted to be able to engage in modern warfare. At the same time it was equally clear that by pushing science, technology, and professionalism in their armed forces they would be risking ideological control, that the more sophisticated their fighting forces became, the less vulnerable they would be to manipulation by the party. The 1989 prodemocracy protests in Tiananmen Square highlighted this dilemma and intensified the debate inasmuch as it was poorly educated troops from the countryside who fired on the demonstrators. Had the troops been recruited from urban centers, trained to operate modern weapons, or otherwise exposed to the skill revolution, some reasoned, they may well have refused to fire on the students. In effect, the potential transformation of the micro parameter has compelled Chinese authorities to face the question of who they deem their enemies to be. As one diplomat put it, they have to decide whether the main threats are posed by the democracy movement and "peaceful evolution" or by foreign invasion:

> Are you going to have a million Soviet soldiers come streaming across the border? Are you going to engage in a high–tech war with Taiwan or the United States? Or are you going to face Tibetans and students? If you perceive your major strategic enemy being inside your own country, then your major concern is for political control.[42]

Furthermore, it can be readily argued that even this dichotomous choice may be short–lived. As the skill revolution spreads to the countryside through global television, satellite dishes, fax machines, and a host of other means, the political reliability of troops from rural areas may diminish accordingly.

For those many countries that are not so preoccupied with political control, the nature of modern warfare is tipping the balance ever more in the direction of sustaining the skill revolution in their armed forces. While eighty-three of the 140 countries with military forces employ some form of conscription, many are now reducing the number of conscripts and the length of their service in favor of recruiting volunteers and professionals.[43] This pattern is partly due to economic constraints that limit the funds available for large militaries, but it would also seem to be very much a consequence of the technical requirements of battlefield readiness.

The implications of these technical requirements for command structures and the maintenance of military discipline are considerable. As the ensuing analysis indicates, the surfacing of authority crises within armed forces can be traced back, in part, to the enlarged capabilities of their personnel. It even seems plausible to conclude that, given the know–how required to operate modern weaponry, the transformation of the micro parameter is likely to have a quicker and more extensive impact upon members of armed forces than any other segment of the citizenry.

The Decline of Discipline and the Erosion of Obedience

Given the refinement of skills at the micro level and the diminished roles, new tasks, global norms, subgroupism, and the many other dislocating constraints that have emerged at the macro level, it is hardly surprising that the transformation of the macro–micro parameter has had major consequences for armed forces everywhere. Again, however, this is not to imply that the numerous signs of authority crises within armed forces and between them and the societies they serve are only a consequence of changes at the other levels. To repeat, the parametric transformations are interactive, with the decline of military discipline and the recasting of command structures serving to reinforce the subgroupism and sharpen the skills of officers and enlisted personnel alike.

It is especially easy to discern changing hierarchical relationships within armed forces and between them and the societies they serve because the traditional command structures of military organizations are so straightforward. At every rank military personnel are supposed to obey the orders of their superiors and, in turn, the defense establishment is supposed to comply with the directives issued by chief executives and their civilian agencies. This ideal of a responsive and responsible military is built into every aspect of military organization, with the wearing of uniforms bearing insignia designating rank and the saluting of those with higher rank being only the more obvious indicators of pervasive hierarchy. To be sure, variability marks the extent to which different armed forces have achieved this ideal, with those in the developed world having been more successful in establishing tight command structures founded on habitual obedience than those in the developing world; but irrespective of the hierarchy achieved in the past by particular countries, all of them appear to be undergoing an erosion of traditional discipline in the present turbulent era.

The authority crises that beset military establishments take several forms. A lesser form, one that may not even escalate into crisis proportions, derives from the fact that no military establishment today can avoid the impact on command structures of its increasing reliance on technical expertise to operate its weaponry. The conventional lines of authority do not follow the distribution of expertise among the ranks. Often highly skilled technicians have to report to less knowledgeable superiors and, as a result, the former may be confronted with situations where they have to ignore their orders and inform their superiors they are wrong. Conversely, frequent may be the occasions when the superiors have no choice but to follow the lead of their more skilled subordinates.

A more obvious and consequential type of authority crisis occurs when an elite officer corps defies constitutional principles and seizes power. While this classic case of military–inspired coups d'état need not detain us here—since it preceded the onset of turbulence and is the subject of a substantial literature—there is one turbulence–generated gap in the coup literature worthy of note: as will be seen, the classic case is likely to occur less and less often. The inhibitions against outright seizures are not so great as to rid world affairs of actual or failed coups—as recent events in Haiti and Venezuela demonstrate—but the advent of parametric change does seem likely to give pause to military elites contemplating illegal takeovers and thus to reduce their frequency.

Yet we do need to examine the more complex and historically atypical forms of authority crisis fostered by the transformation of the macro–micro parameter. Perhaps the most conspicuous of these concerns the capacity of military organizations to maintain and replenish their ranks. Resistance to being drafted by new recruits and desertions by those serving appears to have become more commonplace in recent years. Although in some countries military service is viewed as a way to move out of humble origins (such as among peasants in China), the opposite pattern is more predominant. In most countries, and especially those at war, the proportion of young people who resist and evade being called into service appears to be growing. The figures may not alarm military planners, but the changes are apparently enough for service chiefs in many countries to recommend cutting back on drafts of conscripts and relying more heavily on volunteer personnel.[44] The pattern of draft resistance is also evident in a U.S. Supreme Court decision to shut the door on thousands of asylum claims from young men trying to escape conscription by guerrilla or government forces in strife–torn countries of Central America and Asia.[45] In Germany, a sudden explosion of citizens registered as conscientious objectors occurred as the coalition against Iraq seemed increasingly likely to follow the military option, and a comparable pattern surfaced in Switzerland even though the Swiss make no provision for alternative service, thus compelling resisters to be willing to go to jail. In South Africa, too, draft resistance has intensified as the abolition of the law classifying every citizen according to race led white draftees to question compulsory conscription for whites when the law defining them as white no longer existed. Similarly, when Serbia tried to strengthen its forces against Croatia by mobilizing its reservists, thousands evaded the draft. And even in Japan, which is constitutionally prohibited from assigning its Self–Defense Force (SDF) to U.N. peacekeeping operations and which limits the ranks of the SDF to 300,000, the recruitment target has been difficult to meet despite pay increases, improved work conditions, and better barrack accommodations.

Nor, of course, were the superpowers immune to this form of authority crisis. Its presence was quite noticeable even at those times when they went to war against a clearly defined enemy. Indeed, the spectacle of widespread resistance to service during the U.S. effort in Vietnam and the Soviet campaign in Afghanistan may well have contributed to comparable patterns elsewhere. And as the pattern of draft resistance reached massive proportions in the former Soviet Union in 1989–1990—from 871 in 1987 to 6,667 in 1989 and 135,000 in 1990[46]—so did evading service come to be "seen by young people as normative behaviour, which met with parental approval and found tacit understanding, if not endorsement from Komsomol officials and law enforcement authorities."[47] It has even been said that the Soviets could not even consider direct military participation in the coalition against Iraq because of massive resistance to the idea at home, just as it took less than a week's time for the same resistance to compel the Kremlin to abandon mobilizing reserves in the Russian republic to retake control of Azerbaijan from militants.

As authority crises have hindered the recruitment of military personnel, so have they made it more difficult for defense establishments to maintain their force levels. The readiness of those in enlisted ranks to desert, once an unthinkable idea entertained by a few isolated individuals, has increased noticeably in many parts of the world. Most recruits, of course, complete their tours of duty, but the number who do not is no longer infinitesimal. This is especially so in the armed forces of multiethnic societies, where greater analytic skills and subgroupism nursed by ethnic loyalties has sometimes resulted in extensive defections. When civil war came to Yugoslavia, for example, its Serbian dominated armed forces experienced more than 1,000 desertions by Croats and a quarter of its Slovenian conscripts,[48] a crisis that had devastating consequences for the Serbian campaign against Croatia: "The Croats bent but they didn't break, and in the end it was the army that fell into disarray."[49] Not only did the Serbian army encounter substantial opposition in mobilizing its reservists, but "those who were inducted seemed to have little stomach for battle; Western correspondents have seen army reservists refusing to leave their armored personnel carriers to engage the enemy."[50] In effect, because of the severity of its authority crisis, the Serbian "army's poor results belied its advantage in weaponry and equipment."[51]

As previously noted, somewhat similar developments marked the former Soviet Union when it fragmented into several republics. In this case the pattern was not so much one of desertion as that of troops shifting their allegiances from the central command to that of the republics from whence they came, with the result that the former was no longer in charge of a coherent fighting force. "People are sort of heading for home," one analyst observed. "The question is: what parts are still coherent and which parts are not? The staff arrangement at the center, the strategic nuclear forces are intact, but the ground forces and the military district system are completely undermined."[52] Another analyst described the effects of the authority crisis in the Soviet military even more cogently, noting that the decline of discipline and the erosion of obedience resulting from the inability of the country's national leadership to cope with change had drawn the military inward in a process she called "cocooning":

> They have sequestered themselves from the larger society and are concerned chiefly with hoarding food and supplies to survive the winter.... Officers and enlisted men show loyalty only to their immediate superiors. They respond to their own chain of command. The guy immediately above you is the guy you trust and the guy to whom you report.[53]

It is perhaps noteworthy that tendencies toward disobedience within enlisted ranks are reinforced by the activities of more analytically skillful citizens who pool their efforts on behalf of loved ones in the services. Anxious to protect their sons from the risks of warfare, for instance, parents have mobilized marches and other types of protests designed to exempt their sons from commitments that could risk their lives. Such protests often get considerable publicity, especially if the marchers consist of mothers, and thus add further to the authority crises they are designed to intensify. As the level of fighting increased during the Yugoslavian civil war, for example, officials of the Croatian government "found encouragement in the mothers of army conscripts who traveled by bus from the country...to protest in Belgrade the continued fighting."[54] Likewise, when casualty figures revealed that Montenegrins were paying a disproportionate price in lives, a finding that led to large student rallies and, in one instance, to families of forty Montenegrin soldiers ordered to the front occupying the office of the Montenegrin president demanding the return of their sons.[55] Under special circumstances, moreover, the external incentives toward disobedience may originate with public officials. Authorities in the Baltic republics are a case in point: when the Soviets sought to draft Estonian, Latvian, and Lithuanian youth into the armed service, the fledgling Baltic governments refused to acknowledge the legitimacy of the draft order and helped young people to disobey it. The result was that less than 25 percent of the draftees in each republic actually showed up for duty and in Lithuania the figure was 12.5 percent.[56]

The growing presence of what might be called the "loose cannon" phenomenon is still another indicator of authority crises within the military. This refers to individuals or small groups that decide to take matters into their own hands, defy orders, and engage in acts that have wide consequences. If draft resistance and desertion is, so to speak, the passive form of the erosion of obedience within armed services, the loose cannon is the most active form. Of course, history is not lacking accounts of loose cannons, but it is hardly surprising that their number, variety, and impact appear to be greater with the onset of global turbulence. The very complexity and interdependence of modern societies renders them more vulnerable to a small coterie of individuals who seek to prevent events from evolving in certain directions or who aspire to movement toward different goals. Terrorists are in some ways illustrative in this regard, but they stand out as deviants from societal discipline, whereas military loose cannons deviate from a discipline that does not allow entertaining the idea of individual initiatives. So it is a measure of authority in crisis that accounts of loose cannon seem to surface with increasing frequency. A few examples indicate the breadth of the phenomenon. It was "rogue local units acting without authority" that shelled the quintessentially medieval (and exquisite) city of Dubrovnik in Croatia,[57] just as a U.N.–brokered peace in Yugoslavia nearly came apart when an Air Force jet shot down a helicopter, killing its five European occupants who were monitoring a cease–fire accord, a breakdown of discipline that may have been initiated by a larger coterie of loose cannons than those who undertook the action: as one Western diplomat put it, "The helicopter downing was not the responsibility of two pilots acting alone. The only question is how high up the chain of command the decision was made."[58] Similarly, momentum toward ending a four–year rebellion by an insurgent group in India, the United Liberation Front of Assam, was resisted and slowed down by several military commanders on the grounds that an earlier operation was blunted by the army's sudden withdrawal.[59] And even after peace was successfully negotiated in Angola, there remained a few loose cannons who could undo it: "Now thousands of idle young men without food remain near arms in the assembly points. Without guaranteed food, they will form banditry groups."[60] Even greater, of course is a worldwide fear of loose cannons in the former Soviet Union who have access to stored nuclear weapons and who for any number of reasons may be impelled to use them.[61]

This is not to imply that loose cannons who break ranks ought necessarily to be judged harshly. One can readily imagine situations where their actions affirm one's own values and are thus worthy of applause. It was loose cannon in the national guard, for example, that initiated the process which led to the successful ouster of Georgia's authoritarian president, Zviad K. Gamsakhurdia.[62] Sometimes mutiny in the ranks of the military, in other words, might well be viewed as sacrifice and heroism. The point here is simply that loose cannons, whatever their desirability, reflect the breakdown of traditional lines of authority within armed forces and is thus an indicator of the turbulence to which they are presently exposed.

A final type of authority crisis worth noting concerns the central issue of domestic conflict in which armed forces are ordered to fire on their own people. The issue, of course, is as old as the use of organized coercion, but the historic pattern founded on strict military discipline is one in which militaries have not hesitated to obey the commands to fire. While the prime function of armies has been to wage war against international enemies, the past is scarred by numerous occasions when the line between police and military functions was obfuscated and weapons were turned inward and aimed at domestic enemies. And few, if any, are those historic moments when the orders to shoot on fellow citizens were defied. However, with the advent of turbulence in the parametric controls that sustain the conduct of public affairs, and perhaps especially with the weakening of states and the evolution of global norms that attach high value to human rights, there are solid indicators of a reluctance both to order attacks on fellow citizens and to comply with such orders. The 1989 collapse of the communist regimes in Eastern Europe is a conspicuous instance of this form of authority crisis. The collapse would not have occurred if the militaries in those countries had been ready to comply with orders to use force against their own people. The Ceausescu regime in Romania did issue such orders, but this exception affirms the central tendency. Leaders elsewhere in Eastern Europe, unlike Ceausescu, were aware of the authority crisis at work within their military, and they knew too that their publics would not tolerate such actions (as Ceausescu also found out at the expense of his life).

To be sure, some Soviet forces followed the orders of the coup leaders in August, 1991, and moved on the civilians who sought to protect Yeltsin at the Russian White House. Military support of the coup was short–lived, however, and the central role of the military in that episode was played by the tank battalion that rallied to Yeltsin's defense. But, some might argue, what about the recent roles played by the military in China and Burma? In both cases the army mowed down protesting citizens gathered in public squares and their political leaderships thereby managed to regain and retain political control. Are they not examples that stand out as indicators of the continued effectiveness of strict military discipline? Yes, of course, they are, but again they seem more the exception than the rule (and even in China there were unconfirmed reports that more than 1,400 soldiers and 111 officers above the rank of battalion commander who either fled the scene or defied the order to shoot in Tiananmen Square). Or at least there are enough instances of shifting values with respect to the avoidance of armed force against domestic publics to suggest that military personnel are beginning to set boundaries beyond which their disciplined responses in the command structure are no longer reliable. What has been said about the military in Yugoslavia might well be said about armed services in many other parts of the world: "most of the army realizes they alone cannot hold Yugoslavia together at gun point. The army is only effective as a weapon when it is not used. They recognize that if the Yugoslav national army fires on the people, it loses all authority."[63]

Armed Force and Armed Forces as Agents of Change

If the foregoing analysis is essentially correct and the world is destined to be ensconced in a long, if not permanent, period of turbulence, it seems likely that both armed force and armed forces will undergo increasing diminution as agents of change. The lessened utility of armed force can be anticipated because the world has become too complex to be managed through the exercise of coercive power and the decreasing influence of armed forces can be expected because their command structures have been undermined by the skill revolution and the worldwide crisis in authority. Given the constraints imposed by dense populations, faltering economies, war–weary publics, weakened governments, divided societies, and transnationalized borders—to mention only some of the dynamics that sustain global turbulence—there are fewer and fewer degrees of freedom within which military power can operate as an independent variable that accounts for sizable proportions of the variance in public affairs. High–tech weapons can strike targets with pinpoint accuracy, but neither they nor conventional forces can be very effective when directed toward segments of a bifurcated world marked by shifting loyalties, diverse ethnic groups, decentralized social structures, deep historic commitments, massive migrations, and interdependent economies. And if the personnel charged with exercising the armed force is itself conflicted, uncertain of its tasks, and aware of its limitations, the chances of effective applications of coercive force are further diminished.

The implications of this conclusion for military establishments are profound. They not only suggest that recourse to war across national boundaries is likely to be increasingly eschewed, but they also point to a counterintuitive insight that negates a long–standing and widely shared core premise about the role armed forces play within their own societies: the premise is that the military, having control over the means of violent coercion, is the one institution that can bring order out of chaos and establish societal stability where none existed before. This premise is often voiced by observers fearful that too much societal disorder will lead the military to seize governing power, and it is also a premise that high–ranking officers in many parts of the world have articulated in the hope that their expression of the idea will be experienced as a threat that stabilizes situations bordering on chaos. In some cases, of course, military coups d'état have actually been undertaken on these grounds. As global turbulence mounts, however, the soundness of the premise becomes increasingly questionable. The skill revolution, authority crises, and bifurcated global structures—not to mention population explosions, dynamic technologies, and the globalization of national economies—have rendered societies too complex, resources too scarce, loyalties too diffuse, and military organizations too hesitant for the "military as the only stable institution" scenario to retain its viability. As militaries in Algeria, Argentina, Brazil, Burma, Chile, China, the Philippines, and elsewhere have discovered, and as those in Algeria, Haiti, and Serbia are presently finding, establishing and maintaining societal stability is no longer a simple matter even if military leaders are ready to resort to force and ruthlessly mow down their own people. For many of the reasons noted here, armed forces can no longer assume they have the competence to overcome the dynamics of internal turbulence and set their countries on a stable course; and, of course, the knowledge that this assumption is no longer tenable contributes to their hesitancy to act decisively, a predisposition which, in turn, further lessens their chances of success.

This conclusion lends itself readily to a particular situation presently preoccupying policymakers It suggests that the likelihood of a successful military coup in the former Soviet Union are virtually nil. It could not last long enough to bring order out of disorder even if the military in any or all of the former republics were unified and used all the means at their disposal. The disarray and complexity in that part of the world, along with the potential for opprobrium and countermeasures from other parts of the world, is far too extensive for a military takeover to conform to the historic pattern. To be sure, some Russian officers are "so terrified of the loss of their power, privileges, perquisites and the collapse of everything they believed in and dedicated their lives to, that an effort to seize power cannot be precluded." Yet the probabilities are to the contrary, not only because "very few senior military leaders believe that they can solve the nation's social and economic problems," but also because it is estimated that "the majority fear that any effort to impose order on the country through martial law would be more likely to touch off what the world fears—large–scale civil war in a nation with 27,000 nuclear weapons."[64]

This counterintuitive reasoning that ascribes a growing reluctance to military elites to initiate coups also applies to those situations where a military takeover is not so much a coup against society, but is rather an action that immediately evokes wide support from diverse segments of society. If, for example, persistent economic disarray leads to extensive public backing for the military to restore order and productivity, as appears to be the case today in Venezuela and as many anticipate will occur in one or more of the countries of the former Soviet bloc, that backing is not likely to last long as the intractability of the problems reveals that a military regime is no more able to resolve them than was its civilian predecessor. Furthermore, a widespread public readiness to welcome a military takeover can be misleading: "When soldiers have moved to intervene, people's dissatisfaction with the civilians in power has generally turned out to be not so great after all."[65]

To be sure, most military establishments continue to be more hierarchical than other societal institutions. Except for the current situations in the former Soviet Union and Yugoslavia, the world is not witnessing an open season for the flouting of rank and the breakdown of command structures. On the other hand, traces of undermined discipline and expanded analytic skills fostering redefined self–interests, cocooning, subgroupism, are manifest in many countries, and it is thus self–deceptive to dismiss the most conspicuous situations as mere exceptions, as expressive of unique circumstances that are unlikely to occur elsewhere. For the very same dynamics that brought the Soviet and Yugoslav militaries to their present state are at work on a global scale. The multiethnic foundations of these two situations have doubtless intensified the erosion of their conventional lines of authority, but even the most homogeneous military organization is faced with their hierarchies of rank being undermined by technical expertise and the highly trained personnel required to apply them.

It seems crucial, therefore, to view the Soviet and Yugoslav situations not as exceptions, but as extremes on a continuum, as indicative of a potential to which all armed services are susceptible during a period when established loyalties and organizational practices are everywhere subject to parametric transformations. Soldiers and officers are not exclusively habit–driven; they too are susceptible to learning and change as their skills expand and their opportunities narrow. And the changes they experience are as likely as not to mirror the transformations at work elsewhere in the world. Military establishments and their personnel may be distinguishable by the uniforms they wear and the weapons they shoulder, but otherwise their distinctive missions, roles, and services seem bound to become increasingly blurred by the turbulence that is sweeping across world politics.

Notes

1. *New York Times*, January 10, 1992, A8.
2. *Los Angeles Times*, January 18, 1992, A8.
3. *New York Times*, December 16, 1991, A9.
4. *Los Angeles Times*, January 12, 1992, A8.

5. James N. Rosenau, *Turbulence in World Politics: A Theory of Change and Continuity* (Princeton, N.J.: Princeton University Press, 1990).

6. The analysis which concludes that the system's parameters are undergoing their first profound transformation since 1648 can be found in Rosenau, *Turbulence in World Politics*, chap. 5.

7. For the derivation of this definition of political turbulence, see Rosenau, *Turbulence in World Politics*, 59–78.

8. Rosenau, *Turbulence in World Politics*, 10–11. For a formulation that identifies six parameters, see Mark W. Zacher, "The Decaying Pillars of the Westphalian Temple: Implications for International Order and Governance," in *Governance Without Government: Order and Change in World Politics*, James N. Rosenau and Ernst–Otto Czempiel, eds. (Cambridge: Cambridge University Press, 1992), chap. 3.

9. For lengthy inquiries into both the indicators and sources of the skill revolution, see Rosenau, *Turbulence in World Politics*, chap. 13, and James N. Rosenau, "The Relocation of Authority in a Shrinking World: From Tiananmen Square in Beijing to the Soccer Stadium in Soweto via Parliament Square in Budapest and Wencelas Square in Prague," *Comparative Politics*, 24 (April 1992).

10. Quoted in Rick Atkinson and Gary Lee, "Soviet Army Coming Apart at the Seams," *Washington Post*, November 18, 1990.

11. For an analysis of how weakened governments have operated as a major source of subgroupism, see Rosenau, *Turbulence in World Politics*, chap. 14. The theme that the world has entered the age of subgroupism is elaborated in James N. Rosenau, "Notes on the Servicing of Triumphant Subgroupism," a paper presented at the Conference on the Greek Diaspora in Foreign Policy, sponsored by the Institute of International Relations, Panteois University of Social and Political Sciences, Athens, Greece (May 5, 1990).

12. "The Columbo family civil war...is solid evidence of a widespread phenomenon: the breakdown of autocratic leadership and dissension in the lower ranks.... The system of dictatorial control and unflinching loyalty in all of the families...is fracturing in the same way that the Soviet Union suddenly collapsed." Selwyn Raab, "In the Mafia, Too, a Decline in Standards," *New York Times*, January 19, 1992, sec. 4, 6.

13. A full analysis of the diverse sources of the bifurcation of global structures can be found in Rosenau, *Turbulence in World Politics*, chaps. 10–15.

14. For an explanation of why the terms "sovereignty–free" and sovereignty–bound" seem appropriate to differentiate between state and nonstate actors, see Rosenau, *Turbulence in World Politics*, 36.

15. For a discussion of the problem of labeling the order, see the appendix to my paper, "The New Global Order: Underpinnings and Outcomes."

16. For a discussion of emergent orientations toward the utility of military action, see James N. Rosenau, "A Wherewithal for Revulsion: Notes on the Obsolescence of Interstate War," a paper presented at the Annual Meeting of the American Political Science Association (Washington, D.C., August 30, 1991).

17. James N. Rosenau, "Interdependence and the Simultaneity Puzzle: Notes on the Outbreak of Peace," in *The Long Postwar Peace: Contending Explanations and Projections*, C.W. Kegley, Jr., ed. (New York: HarperCollins Publishers, 1991), 307–328.

18. Melissa Healy, "U.S. Military Will Shelter Haitians at Base in Cuba," *Los Angeles Times*, November 26, 1991, A1.

19. Douglas Jehl and Ronald J. Ostrow, "Pentagon Said to Reject Bigger Anti–Drug Role," *Los Angeles Times*, January 17. 1992, A12.

20. Cf. James N. Rosenau, *The United Nations in a Turbulent World* (Boulder, Colo.: Lynne Rienner Publishers, 1992), chaps. 4–5.

21. It is perhaps a portent of things to come that officers of the Serbian–led Yugoslav army clashed over whether to agree to a U.N. plan to diminish their jurisdiction by introducing peacekeeping troops intended to bring an end to their civil war with Croatia. Cf. Chuck Sudetic, "Yugoslav Army Chief Pledges Support for U.N. Plan," *New York Times*, January 12, 1992, 3.

22. Tim Golden, "Mexican Panel Faults Army in Death of Drug Agents," *New York Times*, December 7, 1991, 3.

23. For a more general assessment indicating that the military are far from alone in having to adjust to the emergence of global norms, see James N. Rosenau, "Normative Challenges in a Turbulent World," *Ethics & International Affairs*, 6 (1992): 1–19.

24. For a cogent analysis of the increasingly central role played by social movements in world politics, see R.B.J. Walker, *One World, Many Worlds: Struggles for a Just World Peace* (Boulder, Colo.: Lynne Rienner Publishers, 1988).

25. William Touhy, "Are Nations Set to Dodge the Draft?" *Los Angeles Times*, August 13, 1991, Sec. H, 4.

26. Roberto Suro, "Female Cadets Charge Abuse on Campus," *New York Times*, October 7, 1991, A8.

27. Larry Rohter, "The Navy Alters Its Training to Curb Sexual Harassment," *New York Times*, June 22, 1992, 1.

28. Serge Schmemann, "5,000 Angry Military Men Gather With Complaints in the Kremlin," *New York Times*, January 18, 1992, 4.

29. For a case of intensive subgroupism on the part of a military unit in the United States, see John Lancaster, "Shrinking The National Guard To Fit the Times: In Mississippi, They're Fighting Mad About It," *Washington Post National Weekly Edition*, February 10–16, 1992, 33.

30. Cf. Rosenau, *Turbulence in World Politics*, 336

31. John M. Broder, "Cutbacks, Criticism Take Toll on Military's Morale," *Los Angeles Times*, January 16, 1992, A18.

32. Ibid.

33. Roger Spiller, quoted in ibid.

34. Elizabeth Shogren, "Officers Demand Unity for Ex–Soviet Military," *Los Angeles Times*, January 18, 1992, 1.

35. Youssef M. Ibrahim, "Algerians, Angry With the Past, Divide Over Their Future," *New York Times*, January 19, 1992, sec. 4, 3.

36. Patrick E. Tyler, "U.S. Concerned That as the Union Breaks Up, So Does the Soviet Military," *New York Times*, December 10, 1991, A9.

37. James F. Clarity, "Ukraine Woos Soviet Troops' Loyalty," *New York Times*, January 10, 1992, A8.

38. Carey Goldberg, "Colonel's Cry: 'It's My Entire Life,'" *Los Angeles Times*, January 10, 1992, A8.

39. Serge Schmemann, "5,000 Angry Military Men Gather," 1.

40. For a discussion of the transferability of computer and other technical skills acquired in the workplace, see Rosenau, *Turbulence in World Politics*, 352–54.

41. Carey Goldberg, "Colonel's Cry."

42. David Holley, "China Debates How to Recruit Today's Army," *Los Angeles Times*, November 12, 1991, H1.

43. William Touhy, "Are Nations Set to Dodge the Draft?"

44. Ibid., 1, 4–5.

45. Linda Greenhouse, "Supreme Court Limits Political Asylum," *New York Times*, January 23, 1992, A7.
46. These figures were released by the Defense Ministry. Cf. Esther B. Fein, "Youths in Estonia Defying the Soviet Military Are Finding Eager Helpers," *New York Times*, April 5, 1990.
47. Natalie Gross, "Youth and the Army in the USSR in the 1980s," *Soviet Studies*, 42 (July 1990), 483.
48. Stephen Engelberg, "Yugoslav Army Trying To Correct Deficiencies," *New York Times*, July 15, 1991, A5.
49. Stephen Engelberg, "Yugoslavia's Breakup," A6.
50. Stephen Engelberg, "Yugoslav Ethnic Hatreds Raise Fears of a War Without an End," *New York Times*, December 23, 1991, A8.
51. Stephen Engelberg, "Yugoslavia's Breakup," A6.
52. Condeleezza Rice, quoted in Patrick E. Tyler, "U.S. Concerned," A9.
53. Ilana Kass, quoted in John M. Broder and Jim Mann, "Military May be Tempted, Experts Say," *Los Angeles Times*, December 10, 1991, A11.
54. John Tagliabue, "Serbia Promises Reply on Truce Plan," *New York Times*, August 31, 1991, 3.
55. David Binder, "Montenegro Balks As Fighting Grows," *New York Times*, November 22, 1991, A4.
56. Craig R. Whitney, "Moscow Is Sending Troops to Baltics to Enforce Draft," *New York Times*, January 8, 1991, A8.
57. Stephen Engelberg, "Yugoslav Ethnic Hatreds," A8.
58. Chuck Sudetic, "Yugoslav Army Chief Pledges Support," 3.
59. Sanjoy Hazarika, "India Reports Insurgents in Assam Agreed to End Their Four-Year Rebellion," *New York Times*, January 16, 1992, A6.
60. Ramiro da Silva, quoted in *New York Times*, December 16, 1991, A9.
61. Elaine Sciolino, "U.S. Report Warns of Risk in Spread of Nuclear Skills," *New York Times*, January 1, 1992, 1.
62. Serge Schmemann, "Stunned, Georgians Reckon the Cost of Independence," *New York Times*, January 10, 1992, A9.
63. Stephen Engelberg, "Yugoslavia: Healing Steps Are Modest," *New York Times*, May 11, 1991, A3.
64. Broder and Mann, "Military May be Tempted," A16.
65. Tim Golden, "Democracy Isn't Always Enough to Repel Attempted Coups," *New York Times*, February 9, 1992, sec. 4, 3.

3

The Future of Transitional Warfare

William R. Thompson

One way to assess the likely role of armed forces in a multicentric world is to evaluate the potential for war in such a system. If the potential for war, or at least war of the major power variety, is as extremely limited as many people believe, the traditional role of major power armed forces is likely to experience considerable change. This conclusion assumes that the existence of major power armed forces has been predicated primarily on meeting threats from other major powers. Eliminate this type of threat and the basic rationale for these forces is undermined. Yet is it safe to assume that the world has changed so much that tendencies manifested in several hundred years of evolution in the major power war system have been radically transformed? I take this to be a theoretical question and it will be to theory that I turn in an attempt to answer it. In particular, I propose to examine six "endism" theories. These are theories about the end of something—ranging from war itself to predictability—that lead to the expectation that the chances of a future major power war are slim to nonexistent.

Major power wars have come in all shapes and sizes, encompassing the continuum from dyadic combat to the global conflagrations in which all major powers are arrayed on one side or the other. The type of major power war on which I will be focusing is referred to here as "transitional warfare." Transitional wars are a shorthand way of designating the most serious type of major power war, the periodic showdowns over which major power(s) will make policy and rules to govern the global political economy. They are wars of transition because an essential precondition for their occurrence is uneven growth and positional movements up and down the pecking order. Without transition, there would be no wars of transition. In the literature, these wars are sometimes called global wars, hegemonic wars, general wars, system wars, or world wars. There is no perfect consensus on which wars qualify for these labels but most would agree that the intramural, major power fighting of 1914–1945, 1792–1815, and 1688–1713 qualify. I would add the 1580–1609 combat as well. The question then is whether we have good reason to anticipate another round of this type of fighting sometime in the twenty–first century. These are the sort of wars which have given major power armed forces their central raison d' être. In the likely absence of such wars, someone will need to invent or emphasize new roles for these forces. Before forecasting the probability of future transitional wars, however, we need to understand why such wars occur in the first place. I turn first, therefore, to a recapitulation of some of the things we know about the causes of transitional warfare. Following that, I review the logic of various endism arguments to see whether they make a strong case for substantially modifying our expectations about the future of transitional war.

Transitional Warfare Assumptions and Dynamics

For the purposes of this analysis, the theory of transitional warfare upon which I will focus is based on leadership long–cycle theory and can be restricted to two sets of dynamics, one global and the other regional.[1] Neither dynamic is viewed as a universal process. Both are conditioned by very specific temporal and geographical parameters. The decade of the 1490s is a convenient starting point for tracing the patterns of structural change that seem most important. The Western European region, in addition, has played a central role as the main arena for the regional dynamics just as European states have long dominated the global processes. The global processes are focused on managing long distance trade problems that became increasingly prominent after the Portuguese found their way to the Indian Ocean. The management of intercontinental questions of political economy depends on the extent to which capabilities of global reach are concentrated in the control of system's lead economy. Historically, one state, the "world power," has emerged from periods of intensive conflict in a position of naval and commercial/industrial preeminence. Naval and later aerospace power, one of the principal manifestations of global reach capability, have been, and continue to be, critical for projecting military force, for protecting commercial sea lanes, and denying extracontinental maneuverability to one's opponents. However, the ability to finance preeminent naval (or areospace) power hinges on a commanding lead in economic innovation and the profits that ensue from pioneering new ways of doing things.[2]

Winning a global war, a period of highly intensive struggle over whose policies will prevail in governing the global political economy, creates literally an unrivaled opportunity for imposing new rules or reinforcing old rules of global behavior. The world power and its allies have defeated the leader's rivals and reduced their capabilities while at the same time the world power has improved its own economic and military capabilities.

Such opportunities come and go. The relative lead in naval power tends to decay. The returns from economic innovation peter out and, frequently, are not replaced by a new surge of creativity, at least in the same place. Old rivals rebuild. New rivals emerge. The coming together of these various tendencies means that the leader's relative edge is gradually whittled away. The postwar concentration of global reach capabilities, the foundations for political leadership and order, is eroded and gives way to a period of deconcentration and deteriorating order.

Within this context, one or more challengers to the world power's leadership will appear. The most dangerous challenger tends to be a state that either has already ascended, or aspires to ascend, to a dominating regional position through the development of its land resources. From the perspective of the incumbent world power, regional hegemony, especially in Europe prior to 1945, is the first step to a more direct attack on the global political economy. From the perspective of the ascending regional power, the world power and its penchant for maintaining the status quo is likely to try to thwart the prospects for further expansion. A clash becomes increasingly probable in order to determine whose preferences will prevail. In the long run, then, global order is intermittent. Phases of relative order are followed by periods of relative disorder, punctuated by bouts of intensive combat to decide what type of order will be imposed. The regional processes of most interest overlap with the central global dynamic. Just as the global political economy has been characterized by successive peaks and troughs of capability concentration, so too has the European regional distribution of power. Regional supremacy meant, among other things, control over adjacent or nearby zones of economic and commercial prosperity. If the global leader lacked some form of insularity from continental expansion, regional supremacy also constituted a direct threat to its continued existence as an autonomous actor. Insularity did not eliminate the threat but only made it less direct. The occasional bids for regional supremacy in Western Europe were always unsuccessful due in large part to the nature of the region's geopolitical configuration. The aspiring regional hegemon would find itself confronted by eastern and western counterweights that were able to inject extraregional resources into the struggle over regional control. The east supplied large armies. The west supplied sea power. Depending on the strategies of the hegemonic aspirant, one or both counterweights were likely to intervene to maintain the balance of power. When both intervened simultaneously, the would–be hegemon found itself fighting a two–front war that it was unlikely to win.

This regional dynamic was unlikely to go on forever without experiencing some modification. Gradually, the ratio of constricted regional capabilities to expanding extraregional resources would become far too asymmetrical to continue encouraging attempts to dominate the European region. The seems to have been a watershed in this respect, although more time may be needed to fully assess whether regional hegemony in Europe is truly an artifact of the past.

Prior to 1945, though, the main regional dynamic was a movement to and from the occasional concentrations of continental power that served as a foundation for attempts at regional domination and threatened directly or indirectly the status quo of the global political economy as well. Combining the central global and regional dynamics suggests that structural change at the two levels were out of phase with one another and yet closely connected. Maritime power concentration was decaying at the global level while continental power concentration at the regional level was increasing. The global erosion no doubt encouraged the ambitions of the ascending regional power. The ambitions of the ascending regional power, similarly, encouraged the reformation of maritime power concentration in order to suppress the regional threat. The forced deconcentration of regional power thus led to, or at least greatly facilitated, the reconcentration of global power. What of the future? Is this regional–global pattern outdated? By mid–1945, it had become clear that Western Europe could no longer play the world's central regional role. The former major powers within the region were exhausted. Two of the strongest major powers outside the region, the United States and the Soviet Union, had been promoted to the rank of superpowers as the major winners of World War II—further demoting the status of the once powerful Western European states. Presumably, the need to be alert to circumstances of global decline and regional ascent had receded significantly.

Yet Western Europe retained considerable centrality in the system as a regional object of contention. A major portion of the Cold War between the United States and the Soviet Union, at least from the American perspective, was focused upon the possibility of Soviet expansion in the European regional theater—very much in the spirit of the historical pattern, although without the immediate prospect of an eastern, continental flanker. That expansion did not occur. If it had, the alignment of regional and global structures might still have resembled the historical pattern, assuming it had occurred in a period of the relative decline of the United States (i.e., certainly not before 1973, if then) and the Soviet Union, at long last, was perceived to be a full–fledged, "European" actor. In this sense, the meaning of the world system's central region could be seen as having expanded, in good Mackinder heartland style, from Western Europe to Eurasia. The other main points of Cold War contention, East and Southeast Asia, could be seen as theaters of the newly expanded, central "region." Other possibilities, now that the Soviet Union has disintegrated and Western Europe has revived, include two conceivable types of expansionary thrusts. One effort might be organized around the ambitions of a central European actor or a unified European actor. Alternatively, another member of the expanded Eurasian region, say a Russia or China, could become the central concern of global power decision makers.

A third possibility is a future characterized by several regions claiming centrality, or at least high significance. The Vietnam War could be said to have been fought in part to discourage the prospects for Chinese expansion in Southeast Asia. The Gulf War was fought in part to discourage the prospects for Iraqi expansion in an oil–rich corner of the Middle East. Neither of these affairs had much potential for becoming global wars because the regions involved were relatively peripheral, particularly in the case of Southeast Asia. The expansionary threats also turned out to be less real than imagined, predicated as sometimes happens on less capability than was actually on hand. Global deconcentration, moreover, had not progressed all that much.

Yet peripheral regions need not remain peripheral forever. Actors once misperceived as expansionary may become genuinely expansionary. For that matter, decision makers have a poor track record in accurately recognizing regional expansion. They have been slow to identify the genuine cases and occasionally exaggerate the threat potential of more dubious cases. What may really matter in the future is the likelihood of bringing together the most powerful regional and global actors

in a situation of intensive conflict and the appropriate set of perceptions about their relative capabilities (declining global and ascending
regional) and respective intentions.

Arguments about the End of Global War

I do not predict that any of these scenarios must take place in the
near or distant future. Whether one of these scenarios does take place
will probably depend on the extent to which the world has changed and
whether the world has changed sufficiently to preclude the various scenarios as future possibilities and perhaps even future probabilities. It is
to this question of the extent of change that I now turn by examining
six endism arguments in the following order: the end of history, the end
of autocracy, the end of war, the end of war profitability, the end of
Westphalian simplicity, and the end of predictability.[3] There is no obvious logical progression; rather there is considerable overlap which suggests the advisability of a general critique as opposed to criticizing
each argument one by one.

The End of History

Francis Fukuyama garnered a great deal of interest in 1989 with a
provocative article entitled "The End of History?" What he had in mind
was the Hegelian notion that history ends when there are no more fundamental contradictions that cannot be resolved within the context of
Western liberalism. This does not mean some form of convergence between rival ideologies nor does it mean an end to ideology per se. Rather,
it means the universal triumph of liberalism, the final and highest form
of ideology, and the exhaustion and rejection of any systematic alternatives. Interestingly, Hegel thought this point had been reached after
Napoleon's victory at Jena in 1806. Fukuyama thinks the point has
been reached or nearly reached in the more developed world only during the past few years.

Fukuyama acknowledges that the universalization of liberalism has
some way to go. Most of the third world remains "very much mired in
history" as he puts it. Neither Russia nor China are liberal societies, he
argues, but there is at least some possibility that their foreign policies
will no longer be driven by Marxist–Leninist doctrine. Hence the
triumph of liberalism is at best an ideational victory, with its full implications yet to be worked out in the real world. For Fukuyama, the im-

plications for the future of world politics are optimistic, if a bit monotonous. He assumes that great power national interests are predicated on ideological foundations. World War I was a battle between liberals and the remnants of absolutism. World War II was fought between liberals and fascists. The Cold War has been a struggle to defeat bolshevism and Marxism. If you remove ideological tensions from world politics, there should be much less chance of large–scale conflict. Large–scale conflict requires large states that are still mired in history. If such dinosaurs should no longer exist—and on this point Fukuyama is prepared to waffle a bit—conflict will persist but it should be small–scale. In the part of the world no longer caught up in history, politics and strategy will cease to hold much interest. Economic strategy, consumerism, and the "Common Marketization" of international relations will prevail as the principal mode of interaction across borders.

The End of Autocracy

The Kantian argument, produced in an earlier end–of–the–century era, has a number of components, several of which intriguingly foreshadow more modern turns in international relations theory.[4] In essence, the argument can be reduced to substituting republican hesitation for monarchical caprice, giving fellow liberals the benefit of the doubt, and avoiding costly interference with commerce. To the extent that conditions develop along these lines, perpetual peace is conceivable. Kant began his elaboration of this position with a strong evolutionary assertion. Republics emerge and survive because they are best equipped to deal with two fundamental political problems: external threats and overly ambitious rulers. These republics are identified in terms of a complex of major institutional characteristics. They must have sovereignty, market economies with private property rights, individual rights, and representative governments with effective legislatures. The development of these institutions produces three by–products. The representation and rights have evolved from a sequence of exchanges between rulers and citizens. Citizens receive some participation in government in exchange for ruler legitimacy, authority, and tax revenues. The last named type of support is particularly critical in market economies. Moreover, the legitimacy and authority of governments facilitate unity among the population, something that is essential to coping with foreign threats. The legal infrastructure associated with the complex of representative government's rights and duties also serve as a restraint on aggressive rulers.

Once republics are firmly established, wars in general are more easily avoided. To the extent that popular consent is necessary before war is declared, the population is likely to serve as a brake on the outbreak of war. War is both dangerous and costly. Lives and wealth must be placed at risk during the fighting. Even worse, the costs of war continue on after the fighting has ended. Wartime destruction must be repaired. Debts contracted during the war must be paid even though it is unlikely they can ever be paid in full as long as the probability of more warfare remains high. Thus it is only natural that republican populations, and their rulers, would hesitate to sanction a war effort.

In decided contrast, nonrepublican rulers interested in engaging in warfare need not seek the permission of their populations. In Kant's view, they also have much less sacrifice to bear:

> war will not force [a nonrepublican ruler] to make the slightest sacrifice so far as his banquet, hunts, pleasure palaces and court festivals are concerned. He can decide on war, without any significant reason, as a kind of amusement, and unconcernedly leave it to the diplomatic corps (who are always ready for such purposes) to justify the war for the sake of propriety.[5]

Yet republican caution and the whims of monarchies are only part of the war avoidance equation. A second factor of some importance is the axiom that, given a world of separate nations and states, the more in agreement people are about their philosophical principles, the more likely they are to assume that accommodation and cooperation are the appropriate strategies to pursue. Apparently, Kant saw this generalization influencing interactions among republican regimes. His explanation centered on the assertion that regimes based on consent are likely to be viewed, by other representative regimes, as similarly organized, just, and therefore to be given the benefit of the doubt when disputes arose. Reminiscent of more contemporary neofunctionalist themes, initial experiences with successful cooperation would make subsequent cooperation more likely. Again, in contrast, republican regimes were likely to suspect nonrepublican regimes which were by definition differently and unjustly organized. As Doyle nicely summarizes this part of the Kantian argument, "fellow liberals benefit from a presumption of amity; nonliberals suffer from a presumption of enmity."[6] Conflicts and wars, as a consequence, are less likely between republican states and more likely between republican and nonrepublican states.

This relationship between regime type and conflict propensity is buttressed further by the observation that as republican regimes become more numerous, the "spirit of commerce" should expand accordingly. As it does, the costs of war and therefore inhibitions against assuming such costs, should also increase. In sum, the elimination of war and the achievement of perpetual peace depend on doing away with nonrepublican regimes. As they disappear from the scene, the reasons for going to war and the likelihood of war decline commensurately.

The End of War

For John Mueller, war has become similar to dueling, foot binding, bear baiting, slavery, lynching and the Spanish Inquisition. All of these practices became obsolete because they were viewed as absurd ways of doing things. They are no longer viewed as acceptable and therefore have become "subrationally unthinkable":

> An idea becomes impossible not when it becomes reprehensible or has been renounced, but when it fails to percolate into one's consciousness as a conceivable option. Thus two somewhat paradoxical conclusions about the avoidance of war can be drawn. On the one hand, peace is likely to be firm when war's repulsiveness and futility are fully evident—as when its horrors are dramatically and inevitably catastrophic. On the other hand, peace is most secure when it gravitates away from conscious rationality to become a subrational, unexamined mental habit. At first, war becomes rationally unthinkable—rejected because it is calculated to be ineffective and/or undesirable. Then it becomes subrationally unthinkable—rejected not because its a bad idea but because it remains subconscious and never comes off as a coherent possibility. Peace in other words can prove to be addictive.[7]

Mueller's analogy is to a person who wishes to descend from the fifth floor of a building. The two main choices are walking down the stairs or jumping out the window. His point is that no rational person would ever consider the second choice to be a real choice. As a consequence, it is unlikely that jumping out of a fifth floor window would ever emerge as a conscious alternative. Why is war like jumping from a fifth floor window? Basically, the answer is that war has become viewed as increasingly ineffective and counterproductive. It accomplishes little but destruction. Therefore, it serves no point. While advocates of this view of war as futile have been around for some time, they had always been in the minority. Now, says Mueller, they are in the majority in the developed world.

Traditionally, a more romantic view of war as a test of heroic valor had prevailed in great power circles. At worst, war was regarded as something that was distasteful but occasionally necessary. A few states, such as Holland, had chosen to drop out of the great power club because they no longer wished to engage in war, but they were the exceptions to the rule that war was a great power way of life.

World War I altered this perception. A repetition of the bloodletting and material destruction was something that had to be avoided at all costs. After 1918, advocates of war as an institution were relegated to a minority position. World War II came about only because of Hitler, Mussolini, and the persistence of *bushido* values in Japan. Hitler, correctly at first, thought he could obtain German goals through risky, incremental manipulation and still avoid a world war. Mussolini thought a new Roman empire could be resurrected in the Mediterranean area at little cost. Japanese values on the utility of war remained unfashionably old fashioned. In their absence, especially in the absence of Adolf Hitler, a second world war might not have even occurred. But by the time it was over, the entire developed world had become "Hollandized." That is to say war in the developed world had become subrationally unthinkable.

Of course, war was somewhat less unthinkable in terms of the East–West Cold War. Fortunately, though, the communists have tended to eschew war as a vehicle for spreading their ideological faith. And even if they were tempted to do so, Western containment policies threatened escalation to World War III as the probable price of territorial expansion. In view of the likely destructiveness of such a war, Mueller suggests that the presence of nuclear weapons may have been a redundant element of Cold War deterrence. The idea that war, or at least major power war, was entirely counterproductive to national goals was the primary inhibiting factor. But then the additional threat of nuclear devastation couldn't hurt.

The End of War's Profitability

Carl Kaysen agrees with Mueller's main conclusion that war is obsolete. He parts company with Mueller on the steps taken to reach the conclusion.[8] From Kaysen's perspective, Mueller has emphasized sociocultural change at the expense of changes in technology, economies, and politics. In doing so, Kaysen thinks Mueller has completely misinterpreted how war's obsolescence came about. Kaysen's starting point is the proposition that wars will not occur if all parties involved see the prospective costs as outweighing the probable benefits. There must be some perceived gains before people will be prepared to go to war. Historically, wars have been profitable for their winners. Fundamental political–economic change since the late eighteenth century, however, has altered the cost–benefit calculus. "Winners," especially in the twentieth century, are no longer likely to profit.

In feudal Europe, the basis of political–economic power in an agrarian economy was the ability to control land and its labor force, as well as being able to convert agricultural surplus into military capability. Victory in war meant more land and more human resources at relatively little cost. The limited scale of the combat meant minimal economic destruction and relatively few people killed. Even if this were not the case, war remained a predominately elite sport that was hardly subject to popular constraints or resistance.

Between the fifteenth and eighteenth centuries, things began to change. The scale of war expanded. Armies became larger. Gunpowder increased the level of firepower attainable. State institutions became more complex. The relative significance of agrarian output compared to urban output and trade began to decline. Yet, since the general level of wartime destructiveness continued to be limited, it was still possible to think war gains outweighed war costs. Although state elites were increasingly unlikely to participate on the battlefield, they still retained most of the power to decide whether wars would occur.

The industrial revolution and its consequences brought about a fundamental transformation of the war cost–benefit calculus. The ways of making things became increasingly mechanized and electrified. Synthetic materials and factories replaced natural materials and handicraft production techniques. Urban populations expanded. Rural populations contracted. Industrialization increased the level of firepower conceivable. Armies grew. Logistical and communication improvements also contributed to the ability to fight on a larger scale than ever before. At the same time, nation–states became more integrated with expanded numbers of people who were literate, middle class, and possessing greater political significance thanks to taxes and ballots. These changes in political economy increased tremendously the costs of war decreased accordingly the possible benefits. Territorial control declined as a significant way to enhance wealth or power. In any event, populations became more attached to their national allegiances which made transfers of political control less legitimate, more costly, and more difficult. The scale and cost of war expanded immensely making it increasingly difficult to recover the wartime sacrifices of life and economic growth. From a rational perspective, it increasingly made more sense to invest national treasure for future productive purposes or to spend it in international exchange then to try to gain wealth through force. It also became increasingly difficult to mobilize consent and support for wars from populations more concerned with other matters than issues of national prestige.

In a sense, Kaysen's argument is about innovations. Innovations in ways of making war and innovations in the ways elite–mass relationships were structured led to innovations in the ways people thought about war. World Wars I and II, rather than serving as Mueller's anachronistic displays, were the catalysts for solidifying the innovations as ways of life. The new ways of warfare were brought home literally and figuratively. The old regimes were swept away and new ones dedicated to popular welfare replaced them. The nuclear weapons introduced at the end of the World War II, far from being redundant, managed to escalate the cost of war even higher. Decision makers became even more cautious. More people became convinced that war could serve no rational purpose. Still, Kaysen does not think that war has become subrationally unthinkable as yet. The attitudinal change is real but it has outpaced equivalent reorientations in the technology of war and the institutions of the state. The Cold War also prolonged thinking about the possibility of war. It continues to be possible to reap political gains from short and inexpensive military victories. Nonetheless, the processes leading to the total obsolescence of war continue and have been facilitated further by the failure of communist ideology. As more nations experience the modernizing consequences of the eighteenth century industrial revolution, the processes should approach completion. However, therein lies the contemporary rub for Kaysen acknowledges that his "fully modern" societies remain a minority in the world system. War is in the process of becoming obsolete but it is not there yet.

The End of Westphalian Simplicity

James Rosenau suggests that international relations scholars who study the causes of war tend to share two biases.[9] One, they are likely to treat the tendency for states to go to war as a constant rather than a variable subject to the possibility of institutional and attitudinal transformations. Two, an examination of war causes tends to exaggerate the weight of war–promoting factors vis-à-vis war–thwarting factors. If one focuses on the obstacles to war, according to Rosenau, it is difficult to escape the conclusion that the probability of interstate war is already low and continuing to decline.

Interstate wars are singled out for special attention because, unlike most forms of internal warfare, conflicts between states are highly visible, unambiguous affairs that demand extensive societal mobilization. In interaction with four sets of processes involving complexity, war weariness, governmental paralysis, and attitudinal revulsion, the war attributes make it more likely that interstate wars are doomed to extinction.

The concept of complexity refers to the increasing density of societal networks that organize groups of people within and across states. These networks can be used to resist efforts to mobilize a society for war; they can also be utilized to resist efforts to conquer and control postwar territorial gains. In general, the "thickening" of societies makes it less likely that outcomes can be controlled by decision makers and, therefore, the probability that military force solutions will be tried should diminish.

Rosenau believes most people are simply tired of interstate violence (war weariness) that costs too much, achieves too little, and diverts attention from more pressing local problems. While there is a well-known hypothesis that war weariness is a cyclical phenomenon, Rosenau argues the current sentiment is not likely to be temporary. In particular, if it leads to political and constitutional prohibitions against military involvement, the aversion to war can become habit forming. In addition to growing complexity and widespread war weariness, governments are said to be paralyzed by their inability to overcome differences of opinion and/or to resolve policy problems. The resulting crisis of national authority structures reduces state legitimacy, encourages popular resistance to national leadership, and makes it more difficult to mobilize for war.

Finally, it is asserted that the increased homage paid to the status of human rights around the world carries with it corollary implications for forms of collective violence. To the extent that state–sanctioned violence is seen as an abuse of human dignity, popular revulsion at the institution of war will be all the greater.

The four processes or dynamics are interactive and mutually reinforcing. None are seen as likely to fade away in the near future. All work to reduce the probability of interstate war. Societies are no longer ready to do battle for remote national goals as the world moves away from a state–centered system toward a multicentric system.[10] As a consequence, national decision makers are left with little room to maneuver their societies into bellicose situations.

The End of Predictability

Robert Jervis's argument evidently is a more general response to John Mearsheimer's argument on the future of Europe.[11] To better understand why Jervis chooses to emphasize certain points over others, some background on the nature of Mearsheimer's contentions should be helpful. Mearsheimer's essay was a theoretically guided speculation on what might happen if the military forces of the United States and the former Soviet Union were to be withdrawn from their respective Western and Central European bases. Essentially, his answer to this question was a prediction of multipolarity and increased instability. This conclusion rested on a number of assumptions. Primarily, Mearsheimer assumed that the distribution and character of military power are the root causes of war and peace. Consequently, the long post–1945 European peace could be attributed to the symmetrical bipolarity of the United States and the Soviet Union and their large nuclear arsenals. Symmetrical bipolarity is a desirable power distribution for four reasons. There is only one conflict dyad that really counts, therefore, the possibilities of war breaking out are fewer. Power imbalances are less likely. So are miscalculations of relative power and resolve. Both factors contribute to the likelihood that deterrence policies will be successful. Finally, the system is likely to be polarized which means that minor powers will have less room to maneuver and manipulate the polar powers into political conflict escalation situations. The presence of nuclear weapons was also viewed from a deterrence perspective. The basic assumption was that deterrence is most likely to be successful when the costs and risks of going to war are great. Nuclear weapons promise horrible destruction if employed. They also suggest the futility of territorial conquest which should discourage temptations toward aggressive expansion.

All of these alleged advantages of nuclear bipolarity would be lost if Soviet and American troops were withdrawn from Europe. Interestingly, the crux of this argument hinges on still another and very curious set of assumptions. Mearsheimer argued that from 1945 on, the Soviet Union and the United States constituted not only the global poles but also the European poles. If they withdrew from Europe, Germany, France, Britain, and possibly Italy would then rise to major power status—presumably on the regional level. Given its proximity, the Soviet Union would also be counted as a major European power. Thus, the European region would have four or five major powers, with or without relatively equal nuclear arsenals, competing in an unstable multipolar setting. Increased conflict and war was therefore more probable in the future than it had been in the recent past.

The easiest criticism to raise would focus on Mearsheimer's curious polar assignments. Does it make sense to treat the United States and the then Soviet Union as European actors? Could not one argue just as easily that Europe's long peace was predicated on the disappearance or at least the eclipse of the indigenous European poles due to defeat and exhaustion in World War II and the ascendance of the two global superpowers. If so, the withdrawal of superpower forces from the European theater might have little impact on the distribution of military power that counts most. The eclipse of the European major powers meant that the regional distribution of power had become secondary to the global distribution. For the regional distribution of power to become more significant one of two things must occur. Either the significance of the global distribution would have to be reduced substantially or the status of one or more European powers would have to be elevated until it equalled, roughly speaking, the status of one of the two global powers. Otherwise, one is forced simply to dismiss the influence of the global power distribution as entirely irrelevant to states' behavior within the region.

Jervis chose a different approach. His argument amounts to a series of reasons why the future of world politics is unlikely to resemble its past. Eight different reasons, based on a mixture of analytical prudence, common sense, and assumptions about the history of international relations, are advanced.

First, students of international relations do not know all that much about how things work in world politics. Few generalizations about how relationships operate have gone uncontested. Forecasting is therefore quite difficult in the absence of a solid theoretical base.

Second, it is unlikely that a single variable will determine the course of events in world politics and even if it has in the past, it may not continue to do so in the future. Jervis cites the example of polarity. While he is dubious that evidence exists to support the contention that bipolarity is always a more stable arrangement than multipolarity, he thinks it may not matter. Whatever its effect was in the past, interaction with other variables in the future may nullify or alter its historical impact.

Third, familiarity with social science findings can influence the way actors behave. Decision makers may even learn not to behave as they have in the past, thereby diluting the predictability of generalizations based on earlier behavior.

Fourth, unless outcomes in world politics are completely determined by external environments, there will always be ample room for the values, preferences, beliefs, and choices of decision makers to play some role. To the extent that foreign policy is strongly influenced by these individual level factors, predicting their future values is a highly dubious undertaking.

Fifth, even if one concedes the dominant influence of the external environment, the current state of affairs is unprecedented. Opportunities for reordering world politics have depended on major wars in the past. And even though the disintegration of the Soviet Union resembles what might have happened if the Soviet Union had lost a war, there is no clear–cut winner in a position to lead the restructuring process. Moreover, despite the weakness of the Cold War's major loser, it remains the only country that could destroy the surviving superpower, the United States. At the same time, the U.S.'s principal allies are also its principal economic rivals. Thus, the uniqueness of the immediate present and near future make it difficult to even categorize the prevailing distribution of power—let alone use it as a basis for prediction.

Sixth is in some respects a more general elaboration of Jervis's second argument. If there is a system of external factors that are tightly interconnected, he argues, then even small changes anywhere in the system can significantly influence relationships elsewhere. Such a system is much too complex to reduce its possible future to the nature of the interaction of two actors, such as the former Soviet Union and the United States. Knowing something about a radical change in the superpower dyad is not sufficient information to be able to map the ramifications for the rest of the system.

Seventh, continuing the point of view expressed in the previous observation, Jervis points out that international relations are sensitive to the influence of chance and accidents. The history of world politics is a sequence of specific events that caused future developments to take a different path than they might otherwise have done. Imagine, for example, the history of the 1919–1945 period in the absence of a first world war in 1914–1918 or the course of the Cold War without a Korean War. Similarly, it is not yet clear whether the Soviet Union's disintegration will be peaceful or bloody, or permanent or temporary. Which way it works out is likely to make some difference to the future of world politics. Hence, contingencies matter and since we do not know what they will be, prediction is exceedingly difficult.

Finally, Jervis employs the familiar metaphors of time's arrow and time's cycles to make his last argument against predictability, which is the most complex and most interesting of the eight. He suggests that analysts tend to interpret macrohistory by relying on one metaphor or the other to define their perspective. Arrow interpretations are based on the assumption that change is gradual, constant, and unidirectional. Cyclical interpretations assume that change moves systems to and from one phase to another without altering the essential patterns. Presumably, both points of view have something to offer, so the main questions is which one has more to offer or provides a better fit and what are the implications of this fact for predicting the future. For example, if arrow interpretations are more appropriate than cyclical interpretations for world politics, predictability will be diminished by new elements introduced by change. In contrast, cyclical interpretations should lead to enhanced predictability because the basic features oscillate rather than mutate. But what if both interpretations have utility? Or what if their relative advantage varies from time to time and place to place?

Jervis contends that the time's arrow metaphor fits the developed world best because it is hard to imagine a war breaking out among developed states. This development is a radical departure from the past and is predicated on three processes: increases in the costs of war, decreases in the benefits of war, and changes in domestic values and regimes.

The costs and benefits of war are viewed largely through nuclear and interdependence lenses. The development of nuclear weapons has magnified the potential negative impact of war exponentially. Interdependence, is so great that no developed state could forgo access to the world economy's flows of capital, investment, and trade. Moreover, the high level of interdependence is made possible in the first place by expectations of a low probability of war. Attitudinally, the idea that a state might become wealthier through territorial conquest than through trade has disappeared completely.

Changes in attitude are at the heart of the emphasis on regime change as well. What Jervis has in mind is that, in the developed world, war is no longer seen as an appropriate technique unless all other options have failed or are blocked. It helps also that there is much less to fight over. In addition to the economic costs, territorial disputes are not particularly salient. Nor is nationalism, and presumably Jervis has Western

Europe primarily in mind here, as potent a factor as it was before World War II. Noting the finding that democratic states rarely, if ever, have fought one another, Jervis points out that developed states are liberal democracies and need not fight each other because they are comfortable with the prevailing status quo.

One might think these changes are sufficiently potent in themselves to make Jervis's case. However, his argument goes much further. It is asserted that the changes in the costs and benefits of war and domestic regimes are very powerful influences on political behavior—more powerful than structural considerations such as polarity. So even if polarity retained some explanatory value in the future, its effects would be overwhelmed by the other types of change. In addition, the cost–benefit calculations and domestic regime factors are not three autonomous variables. They reinforce one another. The high costs of war facilitate higher levels of economic interdependence. Interdependence raises the costs of conflict. Value shifts reduce the perceived benefits of war. Finally, these developments, even though they may have been assisted mightily by perceptions of the Soviet threat and the need to preserve a united front, are irreversible as long as the most developed states retain their democratic values.

These revolutionary changes mean that predictability in the developed world is unlikely.

> [T]hese changes represent time's arrow; international politics among the developed nations will be qualitatively different from what history has made familiar. War and the fear of war have been the dominant motor of politics among nations. The end of war does not mean the end of conflict, of course. Developed states will continue to be rivals in some respects, to jockey for position, and to bargain with each other. Disputes and frictions are likely to be considerable; indeed the shared expectation that they will not lead to fighting will remove some restraints on vituperation. But with no disputes meriting the use of force and with such instruments being in appropriate to the issues at hand, we are in unmapped territory; statesmen and publics will require new perspectives if not new concepts; scholars will have to develop new variables and new theories.[12]

Outside the developed world, Jervis is less optimistic. Time's cycle seems a more appropriate metaphor in Eastern Europe and the third world. Nationalism, ethnic disputes, and regional rivalries are likely but probably containable as security threats to the developed world. The one possible exception is that widespread unrest and conflict in Eastern Europe and Russia might draw German intervention, thereby raising fears concerning German continental dominance once again.

Nevertheless, the possibility of this scenario developing seem quite slight to Jervis.

Endisms and the Future of Transitional Warfare

The six endism theories reviewed above do not exhaust the array of possible arguments. Nevertheless, they are representative and certainly encompass most of the themes: interdependence, economic costs, nuclear destruction, liberalism, democratization, attitudinal and institutional change, and complexity.[13] Taken individually, each theory has variable appeal; taken as a group, the force of the aggregated argument is quite powerful.

Given all these changes, how is it even possible to contemplate the future prospects for major power war? My answer is that, even while acknowledging the power of the forces presumably working against the likelihood of major power war, structural change is also a powerful force. It persists and will continue to persist into the future. Historically, it has been an important promoter of systemic warfare. The question then should not be whether the world is changing. Of course it is. Rather, the question should be are the changes sufficiently powerful to overwhelm the destabilizing consequences of structural change in major power political, military, and economic positions? Ultimately, that is difficult to tell. We lack the theoretical sophistication at this juncture to know how much, say, increasing interdependence is "worth" in terms of persisting uneven growth and positional change. In the interim, though, we can at least reexamine skeptically the asserted obstacles to war. Are the obstacles and arguments that accompany them really as powerful as they seem? One way to evaluate the six sets of arguments about the future of world politics would be to take each one and discuss its individual merits and liabilities. However, such an approach would be both time consuming and redundant. Fortunately, the six arguments share a number of common denominators that make my task easier. It should suffice to extract the common denominators for evaluation. Of course, there are also some idiosyncrasies of the various arguments that deserve and will receive some selective attention as well.

Perhaps the most apparent common denominator of these stories is the shared myth of developmental modernization. In the subfield of comparative politics, it was once argued that political development was the process by which states came to be more like the states that were most advanced, modernized, and not coincidentally, democratic. In other words, as other states in the world system came to resemble the home states of the development analysts, they could then be said to have become more developed. Naturally, there are still analysts who hold this view of political development. But, for many others, the perspective has become an uncomfortable one. It is certainly an arrogant view. How can it be anything but arrogant to say that other societies and polities should become more like one's own as if one's own structures and processes are the only ideal way to organize civic affairs?

The application is not quite the same in the endism literature but it is similar. Essentially, most of these arguments are saying in different ways that more of the world is becoming Western in terms of values, institutions and preferences. Fukuyama stresses liberal ideology. Kant, Doyle, and Jervis emphasize liberal, representative regimes. For Mueller, it is subrationally unthinkable ideas no longer entertained in the West. And Kaysen notes the significance and importance of the ascendance of Western versions of the welfare state. As more of the world becomes Western, major powers become less likely to fight one another. Again, the precise reasons for this reluctance to war against states that are similar vary by argument. Nevertheless, the conclusion is always the same. As the rest of the world becomes more like us, the world will become more harmonious. This generalization may prove to be correct. My initial observation is merely to suggest that we should be careful, if not downright hesitant in our scholarly hubris, in painting a too rosy picture about what the West represents and what a more Western future is likely to resemble. In this case, we are unlikely to be the most objective observers available. Nor should we forget that the modern great power war system was also a Western invention imposed on the rest of the world.

A second, and for our immediate purposes certainly a critical, common denominator is that all of these arguments, whatever their validity, may not be particularly relevant to the future of transitional warfare. All of the arguments, with perhaps Rosenau's as the major exception, recognize explicit spatial limitations on the applicability of their interpretations. They tend to delimit the structural and psychocultural changes in which they are most interested to a subsystem of developed, democratic states in North America, Western Europe, Japan, Australia, and New Zealand. The arguments are always expressed in such a fashion that they relate most directly to interactions and conflict probabilities within that subsystem. Conflict probabilities between states within the subsystem and states outside the subsystem usually are seen as not having changed all that much. Thus one can read most endism arguments as implying that transitional warfare is unlikely within the developed/democratic subsystem but is no less likely between states inside and outside the subsystem. The question then becomes whether there are powerful enough states or states with sufficient growth potential outside the subsystem to cause the type of trouble involved in transitional warfare. Is it possible, in other words, for a nondeveloped democratic state to emerge that could pose at least a major regional challenge? Africa and Latin America seem unlikely sites. But Russia and China, in various possible configurations, remain prominent possibilities in Eurasia. As some of the endism arguments state explicitly, it depends in part on what happens in these states. Regardless of whether that is all that is required, both candidates have a long way to go before they will resemble states in North America or Western Europe.

Allusion to a temporal dimension invokes a third common denominator. Endism arguments tend to emphasize the near future. For example, it is most difficult to imagine another war between Germany and France for control of Western Europe occurring in the next one or two decades. But is it equally difficult to imagine another war between Russia and Germany or China and Japan forty years down the road? The problem is that one must adopt a very long-term perspective to study the prospects of transitional warfare. The near future is too short-term.

At the same time, endism arguments also tend to talk about transitional stages which are nearing completion, but yet incomplete, their fourth common denominator. The world, they say, is becoming more liberal, more democratic, more interdependent, more multicentric, and so on. How long will these transitions take before the probability of war, according to the various arguments, is truly zero? What if the transitional phase takes another one hundred years to complete its transformation? Another one hundred years of a warfare probability greater than zero may represent enough time and threat to caution against assuming the respective transformation is "almost" complete.

A fifth common denominator is that change is seen as irreversible. Democratic states will remain democratic and presumably become even more democratic. The ideological hegemony of liberalism will not be assaulted by the resurgence of old ideas or the invention of new systems of thought. Revulsion from war and its horrific costs will not fade with time and/or new military technology. Levels of economic interdependence will not decline and so on. Yet none of these trends are guaranteed to go on forever.

Most of them are not even likely to be perfectly linear in the first place. The number of democratic states is a good example of this problem. While the trend, over time, is certainly positive, the trend line has been punctuated by major wars, indeed transitional wars, that have done two things.[14] One, they have facilitated the survival of the older democracies in the face of attacks from more authoritarian enemies. Two, they have created circumstances in which either new states emerge with democratic political systems or older states have had democratic political systems imposed upon them. If the other sides had won in the fighting of 1580–1609, 1688–1713, 1792–1815, 1914–1945, there would be far fewer democracies around today. However one chooses to define the meaning of democracy, the majority of states in the system continue to be less than ideal candidates. Presumably, there may still be room for more war–induced democratization just as external pressures continue to contribute to movements in the opposite direction.

The irreversibility problem is even more obvious in the case of Mueller's argument. If World War II can be blamed primarily on Adolf Hitler and Benito Mussolini, as Mueller contends, can we ignore the possibility that environmental deterioration will restore the same type of electoral circumstances somewhere that permitted Hitler and Mussolini to be elected to office? At the very least, it is impossible to argue that it has become increasingly difficult for extreme right and left–wing candidates to garner votes in Western democracies. Nor can we say that with time all democracies have only become more democratic. We might wish otherwise but neither democratization nor economic development, for that matter, are irreversible processes.[15] Yet endism arguments insist on the opposite.

A sixth common denominator shared by these themes is what I would characterize as weak interpretations of previous transitional warfare. By "weak" I mean that their portrayals of these wars, principally World Wars I and II, are not very convincing. Most of the arguments do not bother to differentiate types of warfare other than in terms of their escalating costs. Nor do most of these arguments address the role of structural change in terms of positional rise and decline.[16] Fukuyama views the big twentieth-century wars as primarily ideological in inspiration despite the fact that the winning coalitions in both cases represented a mixture of ideological types. Mueller is even less convincing in his suggestion that World War II might never have happened if Hitler had never been born. Thus, by omission or commission, most endism arguments tend to overlook the possibility that different types of war need different types of explanation and lead to different types of expectations. Different types of war need explanations of varying complexity too. None of the six arguments, for example, operates on the assumption that, historically, one must make a distinction between global and regional theaters of operation. None suggest that "Hollandization" is what happens to states who are squeezed out of the major power ranks because larger, more powerful newcomers have upped the competitive ante (as opposed to decision makers electing to opt out of competition because they no longer wish to participate). Nor do any of the six arguments rest on the assumption that, for a generation or two at least, geopolitical developments outside Europe, in conjunction with wars in Europe and elsewhere, worked to eliminate the transition potential of Western European actors. Who is right? It would only display another type of arrogance to argue that the geohistorical story is superior to stories emphasizing democracy, liberalism, war costs, or war abhorrence. The point is that none of these endism stories appear to have considered the possibility that what they are describing may be due to other factors than the one(s) they choose to stress. Some of the arguments will likely prove to be spurious. At this point in time, it is difficult to say, with any definitive evidence as corroboration, which story or stories is most accurate.

A seventh denominator is not so common but applies particularly to those arguments that incorporate a war cost feature (Jervis's end of predictability, Kaysen's end of war profitability, Mueller's end of war). These arguments bestow an enormous amount of rationality and even omniscience on decision makers who must decide whether to wage war. There can be no disagreement about the impressive escalation in the costs of war over the past several centuries. And given those rising costs, it makes a great deal of sense that decision makers would think twice—and maybe a few more times—before choosing to go to war. However, there are some well–known problems with this perspective. Most notably, decision makers do not always demonstrate a great deal of sense. Given the track record of misperceptions in war decisions, one hesitates to give them the benefit of the doubt.

Another problem is that for most of the participants the costs of transitional warfare in particular have often been greater than the benefits for some time. The major winners are the leaders of the victorious coalition. They get to try to shape the world in their own image. Some of their allies are counted in the winner column despite losing a great deal. The Netherlands and Britain avoided losing their political autonomy to the French and Germans respectively only to lose much of their economic autonomy to one of their coalition partners. Transitional wars also sealed their fates as declining global system leaders. In contrast, the regional challengers invariably lose. With all this repetitive loss, one might think some learning would take place. Between the sixteenth and twentieth centuries this seems not to have been the case.[17] Yet perhaps this observation hints at another side of the coin of war cost–benefit calculations in transitional wars. War costs are frequently indexed in terms of lives lost, property destroyed, and resources consumed. War benefits are more traditionally restricted to territorial gains. Territorial gains and losses are part of transitional warfare too—but only a part and primarily of interest to expanding regional challengers. The cost–benefit calculation becomes even more problematic if we add the opportunity to structure the global political economy. How much is it worth to "rule the world?" How much is it worth to make sure the other side does not rule? At root, that is what transitional wars are about or become if they do not start that way. If the past is any judge, this is a benefit for one side and a cost for the other that is worth quite a bit.

Even so, the huge stakes that emerge in these contests must be weighed against the previously noted misperceptions of decision makers about the probable duration of war and the identity of their adversaries. Decision makers too often expect short wars just as they tend to overestimate their own capabilities and underestimate those of their opponents. If we take these decision–maker pathologies, which may or may not have been influenced by the presence of nuclear weapons, into account along with the great stakes involved in transitional warfare, we should be extremely cautious about the straight forwardness of the cost–benefit calculations likely to be associated with these infrequent bouts of global bloodletting.

Does that leave Jervis's more general unpredictability argument as the safest approach to adopt? That course will no doubt appeal to some, but there are several points that need to be made about the Jervis position. First, we will all agree that there are limits to the predictability of the future. Just because certain types of wars have come about in a consistent fashion for several centuries does not mean that they must continue to do so in the future. Transformational change is conceivable. Practices, institutions, and attitudes are susceptible to change. The ambiguities surrounding the probability of cyclical processes persisting create their own sense of unpredictability.

However, the argument that the future is entirely unpredictable will be most appealing to analysts who have consistently maintained that prediction in world politics is difficult and unlikely. That is to say one needs to be careful in differentiating between perennial factors working against predictability and newly erected barriers. In Jervis's case, his position on this question has been fairly consistent before and after recent developments. Some examples should help to support this observation.

In the late 1960s, Jervis wrote as part of an evaluation of the use of quantitative techniques in international relations:

> Most quantitative studies have not successfully provided more than correlations.... there are several problems which may mean that statistical correlation is not the best path to understanding in this area. Simple correlations will not appear if the effect of one variable depends in part on the state of other variables. In many cases one factor can contribute to different, and even opposite, effects, depending on the circumstances, and one result can be produced by very different combinations of variables.[18]

In the mid–1970s, Jervis advanced a number of caveats in his own review of misperceptions including the following:

> This is not to claim that we will be able to explain nearly all state behavior. As we will discuss in the context of learning from history, propositions about both the causes and the effects of images can only be probabilistic. There are too many variables at work to claim more. In the cases in which we are interested, decision-makers are faced with a large number of competing values, highly complex situations, and very ambiguous information. The possibilities and reasons for misperceptions and disagreements are legion. For these reasons, generalizations in this area are difficult to develop, exceptions are common, and in many instances the outcomes will be influenced by factors that, from the standpoint of most theories, must be considered accidental. Important perceptual predispositions can be discovered, but often they will not be controlling.[19]

The point is not that Jervis has always been mindful of the inherent difficulties in controlling for intervening variables and theorizing parsimoniously. That is a fact that need not be disputed. Rather the point is that much of Jervis's argument in the early 1990s resembles points that he advanced in the 1960s and 1970s. Presumably, this is more a function of his general skepticism about explanation and prediction and not necessarily a function of new developments in world politics.

Thus much of Jervis's argument betrays an ingrained skepticism as opposed to a new found one. If many of his reservations about predictability preceded more recent developments, they probably could have been made in earlier centuries just as easily. Yet despite these reservations, strong structural patterns have emerged from the battle–scarred history of the last half millennium.

But what should we make of his belief that the significance of structural variables such as polarity have been eclipsed by war cost–benefit calculations and domestic value changes? Unfortunately, that is all that it is at this juncture, a belief. Jervis presents neither evidence nor much argument that this is the case. He simply stipulates it.[20] Something more concrete would be preferable before we heed his radical advice to "develop new variables and new theories," especially in a period of flux and especially if some of the old variables and theories possibly retain utility.

Conclusion

In conclusion, these various endism arguments give too much credit to time's arrow and not enough to time's cycles. Indeed, these arguments yield exclusively to time's arrow, giving no credit to time's cycles. The closest we come to a cyclical argument is Jervis's attribution of cyclical characteristics to Mearsheimer's argument about bipolarity and nuclear weapons. But this is not really much of a cyclical argument. Most realism arguments are really flat arrows; change is minimal and behavior does not oscillate back and forth between phases.

Clearly, we should strive to avoid the analytical trap of being forced to choose between these arrow and cycle alternatives. Gould's advice on this problem seems most appropriate.[21] When asked which alternative he prefers he answers "both and neither." Both types of interpretation are capable of improving our understanding. Neither type of interpretation is likely to be sufficient to the exclusion of elements from the other end of the continuum. We need integrated interpretations that combine arrow and cycle. The long–cycle argument contains elements of both types of change.[22] But when you combine the transitional arrows of endism and the historical–structural cycles of world politics and economics, you cannot yet reach the sanguine conclusion that only the arrows count and that the cycles have lost their significance. The cyclical element appears to remain too strong and the arrows of endism not yet strong enough. We may wish it were otherwise but it will probably take more time to eliminate the legacies of the past. Until that happens, we do not yet have enough theoretical reason to anticipate that the future of transitional warfare will stray too far from the trodden path of its own destructive history.

This conclusion does not mean that future transitional warfare among the major powers is inevitable—only that its demise within the next generation or two is less than a sure thing. Indeed, if we assume that it cannot happen again we are more likely to facilitate its reoccurrence than if we assume that, unfortunately, it remains very much a possibility.

Notes

1. See George Modelski, *Long Cycles in World Politics* (London: Macmillan, 1987); George Modelski and William R. Thompson, *Seapower in Global Politics, 1494–1993* (London: Macmillan, 1988); William R. Thompson, *On Global War: Historical–Structural Approaches to World Politics* (Columbia: University of South Carolina Press, 1988); William R. Thompson, "Dehio, Long Cycles and the Geohistorical Context of Structural Transitions," *World Politics* 43 (1992); Karen Rasler and William R. Thompson, "Concentration, Polarity and Transitional Warfare," paper presented at the annual meeting of the International Studies Association, Atlanta, Georgia, April 1992.
2. George Modelski and William R. Thompson, "K–Waves, the Evolving World Economy, and World Politics: The Problem of Coordination," paper presented at the annual meeting of the International Studies Association, Atlanta, Georgia, April, 1992.
3. The "endism" term is taken from Samuel Huntington, "No Exit—The Errors of Endism," *The National Interest* 17 (1989): 3–11. Huntington looks at some of the same arguments but from a much different theoretical stance. My employment of the endism term is also meant to be more neutral than was the case in the Huntington article. The six perspectives I examine do not exhaust the field of possible explanations. See, for example, J. David Singer, "Peace in the Global System: Displacement, Interregnum or Transformation?" in *The Long Postwar Peace: Contending Explanations and Projections*, Charles W. Kegley, Jr., ed. (New York: HarperCollins, 1991). Singer lists twelve alternative hypotheses on the post–1945 "long peace." See also Jerry W. Sanders, "The War of Historical Interpretation and the Prospects For Peace In the Post–Cold War Era," in *Research in Social Movements, Conflicts and Change*, Louis Kriesberg and David Segal, eds. (Greenwich, Conn.: JAI Press, 1992). Sanders discusses some interpretations not reviewed here. For an emphasis on the role of norms in ending war, see James L. Ray, "The Abolition of Slavery and the End of International War," *International Organization* 43 (1989): 405–439 and James L. Ray, "The Future of International War." Paper delivered at the annual meeting of the American Political Science Association, Washington D.C., August 1991.
4. Immanuel Kant, *Kant's Political Writings*, Hans Reiss, ed., H.B. Nisbet, trans. (Cambridge: Cambridge University Press, 1970). My review of the Kantian argument relies heavily on Michael W. Doyle, "Liberalism and World Politics," *American Political Science Review* 80 (1986): 1151–69.
5. Quoted in Michael W. Doyle, "Liberalism and World Politics." *American Political Science Review* 80 (1986): 1161.
6. Ibid.
7. John Mueller, *Retreat from Doomsday: The Obsolescence of Major War* (New York: Basic Books, 1989), 240.
8. Carl Kaysen, "Is War Obsolete? A Review Essay," *International Security* 14 (1990): 42–69.
9. James N. Rosenau, "A Wherewithal for Revulsion: Notes on the Obsolescence of Interstate War," paper presented at the Conference on Peace and Conflict in International Relations, Peace Research Institute, Frankfurt, Germany, May 1992.
10. See James N. Rosenau, *Turbulence in World Politics: A Theory of Change and Continuity* (Princeton, N.J.: Princeton University Press, 1990).
11. Robert Jervis, "The Future of World Politics: Will It Resemble the Past?" *International Security* 16 (1991/92): 39–73; cf. John J. Mearsheimer, "Back to the

Future: Instability in Europe After the Cold War," *International Security* 15 (1990): 5–56.

12. Robert Jervis, "The Future of World Politics," 55.

13. For instance, readers may not need reminding that many of these arguments have been made before. Kant wrote in the late eighteenth century. Hegel first pronounced the end of history in the early nineteenth century. Manchester liberals stressed the pacifying effects of commercial interdependence in the mid–nineteenth century. On this point see Geoffrey Blainey, *The Causes of Wars* (New York: The Free Press, 1973). Norman Angell, in his *The Great Illusion: A Study of the Relation of Military Power to National Advantage* (London: Heinemann, 1914), emphasized the prohibitive escalation in war costs. And Jean de Bloch, in his *The Future of War* (Boston: World Peace Foundation, 1914), predicted that changes in military technology and strategy had made war impossible.

14. George Modelski and G. Perry, "Democratization in Long Perspective," *Technological Forecasting and Social Change* 39 (1991): 23–34.

15. We should not even assume that as obnoxious a practice as slavery is irreversible. For that matter, it is not all that clear that slavery and its functional equivalents have been as effectively eliminated as we might like to think. It should also be noted that the elimination of slavery that has been accomplished required armed force (British naval patrols) and war (the American Civil War) among other factors. See David Eltis, *Economic Growth and the Ending of the Transatlantic Slave Trade* (New York: Oxford University Press, 1987), for a good discussion of the complicated processes surrounding the nineteenth century suppression of the slave trade.

16. James N. Rosenau is the one partial exception. In his paper "A Wherewithal for Revulsion: Notes on the Obsolescence of Interstate War," presented at the Conference on Peace and Conflict in International Relations, Peace Research Institute, Frankfurt, Germany, May 1992, he mentions the traditional conflict inhibiting role of hegemons but he assumes the future will be too complex to produce hegemonies (another endism?). Even if such a concentration of power were to emerge, its decision makers would be constrained from using military force to resist challenges.

17. See Russell E. Weigley, *The Age of Battles: The Quest for Decisive Warfare from Breitenfeld to Waterloo* (Bloomington: Indiana University Press, 1991) for a recent argument that between 1631 and 1815, in an era thought to be conducive to effective and decisive warfare, there were actually few clear winners and losers in European warfare.

18. Robert Jervis, "The Costs of the Quantitative Study of International Relations," in *Contending Approaches to International Politics*, Klaus Knorr and James N. Rosenau, eds. (Princeton, N.J.: Princeton University Press, 1969), 212.

19. Robert Jervis, *Perception and Misperception in International Politics* (Princeton, N.J.: Princeton University Press, 1976), 31.

20. Curiously, no endism argument reviewed here presents systematic evidence of changes in mass or elite attitudes toward war.

21. Stephen Jay Gould, *Time's Arrow, Time's Cycle: Myth and Metaphor in the Discovery of Geological Time* (Cambridge, Mass.: Harvard University Press, 1987).

22. For example, George Modelski, "Is World Politics Evolutionary Learning?" *International Organization* 44 (1990): 1–24.

4

The Shifting Threat and American National Security: Sources and Consequences of Change

Donald M. Snow

The dramatic, unprecedented, and largely unforeseen end of the Cold War has profoundly and adversely affected the intellectual foundations of American national security policy. The constructs based on an apparently immutable East–West military and political competition have been overtaken by events. Whether the reason was the triumph of the West and the vindication of the American strategy of containment is not the issue here. Certainly George F. Kennan's entreaties about the long–term outcome of communism seem more prophetic now than a generation ago, but this "success" leaves us with a void: on what basis will American security policy be formulated for the remainder of the century and beyond?

Security policy is that part of foreign policy that deals with threats to the national interest.[1] In its most traditional sense, security policy addresses those interests that are threatened by the hostile will of opponents who ground that hostility in armed force. Although that definition of security almost certainly will be broadened in the future (if it has not already been), this notion of security clearly underlay the competition with the former Soviet Union. As an opponent, the Soviets met both criteria of a traditional national security concern: the ideological evangelism and/or geopolitical ambitions of the Soviet Union clearly opposed American vital (and not so vital) interests, and the lingua franca of Soviet power was military, especially in Europe.

American national security policy since it was clear in the latter 1940s that the wartime collaboration with Soviets could not continue has been grounded in countering that threat. America's most vital interests centered on a continuing free and prosperous Western Europe and Northeast Asia, and the status of vital interest was extended to the Persian Gulf by President Carter in 1980. In each case, communism provided a plausible opponent, and in the 1960s, the Soviets extended the competition to other parts of the third world, where the vitality of interests to either party was not always evident.[2]

The pattern of American interests has not changed. A free and prosperous Europe to which the United States maintains commercial access remains vital to American well–being; Japan remains a vital partner despite the current wave of "Japan–bashing"; and the Persian Gulf War of 1991 offers obvious testimony of the continuing relevance of access to Persian Gulf oil.[3]

What has changed is the structure of threats to those interests. In national security terms, the balkanization of the former Soviet Union into a series of increasingly contentious sovereign republics has meant the dissolution of the old threat structure. The threats to European tranquillity are now largely internal as nationalism is unleashed in the former communist world that stifled national self–expression; the violent disintegration of Yugoslavia, Hungarian irredentism in Romanian Transylvania, and religious/national Armenian–Azerbaijani venom in Nagorno–Karabakh are examples. Only rumblings of North Korean nuclear aspirations—accentuated by that country's announcement in 1993 that it would leave the Nuclear Nonproliferation Treaty (NPT) regime—mar the peace in northeast Asia; Desert Storm has for the time stifled threats to Persian Gulf oil.[4]

The result is what might be called an *interest–threat mismatch* and accompanying debate over the direction American national security strategy should take. The idea of a mismatch is straightforward if confounding and consists of two propositions. First, in those physical areas of the world where American interests are most important, there are decreasingly any military threats of consequence. Second, the concentration of military threats, in the form of latent or actual instability and violence, is in those parts of the world where American interests are generally less than vital. Given that vital interests are conventionally viewed as those worth going to war over, national security calculations based on military terms are consequently brought into question.

Because they represent radical departures from the past, each proposition requires justification. The chief menace to America's obviously vital political and economic interests in Europe, Asia, and the Persian

Gulf was always the Soviets or their proxies. This was most clearly the case in Europe where, noncoincidentally, the United States mustered its greatest military efforts. The Soviet threat has, however, disappeared. At the most obvious level, there is no longer a Soviet Union to produce such an effect. Although the former Soviet nuclear force remains largely intact and thus a physical problem, that is likely to be decreasingly a problem as arms control efforts reduce nuclear arsenals. The major nuclear threats are internal to the former Soviet Union in the possession of theater weapons by the successor republics, who may sell them to nuclear aspirants (as reports of sales to Iran in March 1992 suggest) or in their threat or use against other former Soviet rival republics.[5]

Soviet offensive conventional power has simply disappeared. In the early days after the union dissolved into the Commonwealth of Independent States (CIS), there was brave talk about maintaining a common military structure; first the Ukrainians and now the Russians have jettisoned that idea in favor of national armed forces. The result is a fragmentation of former Soviet forces among the republics; no single unit poses an offensive threat, and even if the parts could be aggregated into a new whole, it would be no more than a shadow of the old whole. For the foreseeable future, former Soviet forces will be incapable of serious projection outside the boundaries of the old union. Those forces now pose a threat to one another, not to anyone else. Their most plausible uses will be to protect suppressed minorities—especially Russians—residing outside their "native" republics. They also likely will become part of future U.N. peacekeeping efforts.

The opportunities for military activity reside overwhelmingly in the third world, especially Africa, Asia, and the Middle East. In this part of the world, American vital interests are almost exclusively in resource access. Materials science is on the verge of producing its equivalent of genetic engineering in the form of molecular rearrangement that will allow the production of materials with any desired characteristics from common elements, reducing dependence on many mineral resources. That leaves petroleum as the only third world commodity clearly worth fighting over. Beyond that, "it is clear that no single conflict in the third world poses an explicit threat to U.S. national security."[6]

This mismatch is totally unlike the Cold War. For forty years, there has been a neat symmetry: our vital interests and what threatened them coincided. Occasionally, as in Southeast Asia, a threat might appear that was not unambiguously vital to throw the construct out of kilter, but the centers of interest and threat vitality clearly focused on the Soviet Union. Interest and threat coincided.

In the post–Cold War world, interests and threats hardly overlap, much less coincide, a situation "where strategic stakes are modest or absent."[7] The result is fluidity in thinking about strategy which founders on the attempt to recreate the overlap. There is resuscitated concern about a North Korea more likely to merge politically with the South than to attack it (what rationale could possibly underlay destroying a South Korean economy that could make North Koreans prosperous?), and veiled reference to another outbreak in the perpetually unstable Persian Gulf.

Increasingly, the identified enemy is *uncertainty*, the outbreak of "unforeseen and unforeseeable" conflicts threatening American interest, to borrow language from General Colin S. Powell, chairman of the Joint Chiefs of Staff, in February 1991 testimony before the Senate Armed Services Committee.[8] Uncertainty may in fact be the problem, but it is a tough opponent to deal with. Since it cannot be specified (by definition) in detail, it is hard to sell to a penurious Congress and the American people. Moreover, most of the uncertainties lie outside obvious American interests, leaving the criticism of "who cares?" about these problems. Uncertainty is difficult to plan against, unless one takes the fiscally untenable position that one must consequently plan and prepare for everything and where "the United States probably will not enjoy the certainty of having either permanent friends or permanent enemies in the third world."[9]

This debate is just beginning to emerge. At the end of the debate, hopefully, will be a new consensus on the nature of the threat and what to do about it. The outcome of that debate will have enormous implications for American national security policy and the strategy and forces that implement that policy.

Sources of Change

One of the most remarkable aspects of the end of the Cold War was how poorly it was anticipated. Only months before the edifice of communism began to crumble, Americans were still debating things like nuclear force modernization in Europe (the Lance missile). Historians will have a field day unraveling retrospectively the failure of accurate prediction, and it is only with the virtue of hindsight that we can begin to unravel the web. Doing so is necessary to understand where we are today.

The discussion can usefully be organized around two constructs. On the one hand are a series of geopolitical events that help define current reality. The major geopolitical phenomena are the revolutions of 1989 and beyond, the weakening and ultimate collapse of the Soviet Union as a politico–military entity, and the American military triumph against Iraq. On the other hand, a series of emerging trends that have their roots in the 1980s play an important role in defining that reality. These include the effects of the technological or third industrial revolution on the international system, the telecommunications revolution, and a broadening conception of the basis of national security.

Geopolitical Events

What may well come to be known as the Gorbachev Transition in international politics saw seismic changes that dismantled the structures and rules of the post–World War II system and laid open the way for a new set of relations for the post–Cold War world. Gorbachev almost certainly did not wish all of the things to happen that transpired, but he did unleash forces that spun out of his control.

Gorbachev's most notable legacy was the destruction of that part of the communist world that posed a security threat to the Western world, the Warsaw Treaty Organization (WTO). This demolition must be viewed in three separate steps: the loosening of Soviet control over the countries of Eastern Europe that permitted the revolutions of 1989 to occur; the formal dismantling of the WTO in 1991; and the dismemberment of the Soviet Union during 1992.

Gorbachev's decision to loosen Soviet control of Eastern Europe set the process in motion. In retrospect, the reasons are as clear as they can be until Gorbachev's memoirs are available: the WTO countries required economic modernization; the restive populations of the WTO countries chafed at standards of living noticeably lower than their Western European counterparts; governmental corruption and inefficiency were increasingly public and embarrassing; and the continued "enslavement" of Eastern Europe was the most visible symbol of East–West differences. In these circumstances, Gorbachev and Foreign Minister Eduard Shevardnadze traveled the WTO area in summer 1989 with a simple message: "Make peace with your people; there will be no Soviet tanks to save you this time."[10]

In all likelihood, Gorbachev and those around him had little idea of the intensity of animosity toward communist rule in Eastern Europe and thus the intensity and speed with which populations would shunt aside communism and the communists.[11] By the end of 1990, hardly anyone dared admit to being a communist; by the end of 1991, the only communist governments left were in remote places such as North Korea, Cuba, and China.

The fall of communism meant the end of the ideological competition that had conceptually underlay the East–West military competition. As the former WTO states—especially the "northern tier" states of Poland, Czechoslovakia, and Hungary—moved toward market democracy and the former East Germany became part of the Federal Republic, there was nothing left to fight about. The formal dissolution of the WTO in spring 1991 was just a formality.

The death of the WTO was the first giant step in disentangling the military confrontation. It meant not only that the Soviet Union no longer had a buffer zone to protect it from an invasion from the West. More importantly, it meant that the Red Army would have to retreat from Eastern Europe back to Soviet soil (a process not yet complete). In that circumstance, the Soviets could only attack westward by first attacking and fighting their way across the territory of their former allies, thereby greatly complicating the military problem for them.

Even the problematic prospect of fighting across Eastern Europe became academic with the second geopolitical event: the physical dissolution of the Soviet Union. This dismemberment, which took place at the literal end of 1991, was unprecedented in international history. Nation–states and empires (the Soviet Union qualified as both) have broken apart before, but never peacefully and by agreement not coercively based. It is a political phenomenon that will be studied for years, if only because of its precedential possibilities as a number of third world countries approach democratization. In the current context, however, it is the military implications that are most important.

The basis of the Soviet claim to superpower, even major power, status was always grounded exclusively in military power. As a military power set aside from all others but the United States, its great thermonuclear arsenal defined its status as superpower. For the time being, that defining characteristic still holds, although jurisdictional wrangling among those republics—Russia, Belorus, Ukraine, and Kazakhstan—who maintain control over the strategic arsenal remains a problem. Boris Yeltsin, president of the Russian Republic that controls the largest part of the arsenal, declared in early 1992 that the United States was not his enemy and thus he no longer targeted the U.S. with those weapons.

That may be correct, but thanks to the wonders of modern electronics, it is also a situation that could be changed very quickly. Once jurisdictions are worked out, arms control processes will almost certainly pare down the arsenals, but likely to levels that leave the Americans and the Russians clearly superior to any other nuclear pretenders.[12]

It is on the conventional side that change has been most dramatic. Even before disintegration, Soviet forces were in disarray, with record desertions, draft dodging, and reports of deteriorated morale. Now, the Soviet armed forces as such no longer exist. The attempt to maintain central control through the CIS was always wishful thinking given the ferocity of nationalist sentiments. In March 1992 Russia, in a move soon followed by the other republics, announced it would form its own army; the announcement simply nailed the lid on the coffin of the Soviets.

What then is left of Soviet conventional military might? There is a great deal of equipment still around, although much of it is up for sale, and there are jurisdictional disagreements over who owns fundamental units, such as the Black Sea fleet. Within the services, soldiers of the former Soviet armed forces are dropping their insignias and adopting those of their new republics. Somewhere down the road those forces may be reconstituted as a formidable array of forces. For the foreseeable future, that will not be the case. It is not too strong a statement to say that for the remainder of the century, the forces of the former Soviet Union will be incapable of projecting offensive force outside the boundaries of the former Soviet Union. The military aspect of the Cold War is truly over.

The other defining geopolitical event is the Persian Gulf War. It is important because it was the first major military event of the post–Cold War world and because it completed the renaissance of the American military from the Vietnam embarrassment. The "lessons" of Desert Storm are numerous and open to debate. For present purposes, however, I will identify and discuss three legacies that are not entirely obvious but which may be of the greatest influence on the new international order.

The Persian Gulf War is already one of the most studied and analyzed in human history, despite its short duration and the absence of spirited competition. The professional American military system of highly trained and motivated professional soldiers and evolved strategies and tactics have rightly been praised for their performance.[13] The Americans, as Taylor and Blackwell assert, "had better personnel, better ideas, better technology, and better equipment."[14] The conditioning rejoinders of ideal physical conditions (the desert) unlikely to be encountered again and an extremely inept opponent provide limits on extrapolating the experience into the future.

Three outcomes of the war are likely to be instructive in the future. The first has to do with the enormous gap between the military capabilities of the first world—notably the United States—and the countries of the third world. As a number of observers have noted, the Western powers had superior forces in terms of training and motivation,[15] qualitatively superior equipment which those troops could utilize to maximum effect (something not true of earlier high–tech weapons), and superior organizing ideas to utilize their superiority to maximum effect strategically (the "offset strategy") and tactically (high–maneuver warfare). The Iraqi forces were simply outmanned, outgunned, and outthought.[16]

It must be remembered that the Iraqis were thought to be a formidable opponent who could exact high American and coalition casualties as the price for liberating Kuwait. The Iraqi army was the fourth largest in the world, and it was battle–tested in the long war with Iran. What got lost was that it was the veteran of a war with a similar third world country. Against a modern, professional Western force, the Iraqi conscript army did not stand a chance. The ease with which the U.S. and its partners dispatched Iraq provides a first legacy of the war: the capabilities gap between the most advanced Western military machines and third world militaries is so vast—and growing—that third world states are almost certain to avoid conflict which might draw the West—especially the United States—in opposition on anything like American terms. From a third world vantage point, this conclusion seems inexorable.

This conclusion could be reversed in one or more of three ways. First, third world countries could narrow the gap. This, however, is unlikely. On the one hand, the gap would have to be narrowed by acquiring the most advanced Western technologies, which likely will be restricted in the future. On the other hand, third world militaries are notoriously poor at maintaining or independently operating the most sophisticated weapons. Second, the United States, out of budgetary calculations, could dismantle or seriously degrade its own capabilities, a contingency against which the military would object strenuously. Third, third world countries could conclude that first world power would not be brought to bear against them in specific situation, a calculation that could be correct or incorrect.

A second legacy has to do with the precedential nature of Desert Storm both for future coalition warfare and for collective security. The coalition was impressive both for the size of its membership—nearly thirty nation–states—and its diversity. Most notably, the coalition brought together both Christian first world nations and a diverse range of Islamic, including Arab, states against the Arab state of Iraq. The Americans, French, and British provided the bulk of the actual forces employed, of course, and some participation from the Islamic world was more symbolic than anything else (e.g., Bangladesh). At the same time, the question of whether this symbolized the wave of a future of broad–based opposition to aggression was raised as a new world order possibility.

One must be cautious in any assessment. Arab participation in the effort, after all, arose because Saddam Hussein violated a basic, if implicit, rule of the Arab world: Arab states do not alter the 1919 boundaries of the region by force, which he did by conquering and annexing Kuwait.[17] The Arab members of the coalition joined to reverse this wrong; had the coalition purpose been expanded to punish or overthrow Saddam by invading and occupying all or part of Iraq, the coalition would have dissolved instantly. Desert Storm is also unique because it was the first instance where the employment of force was authorized under the collective security provisions of Chapter VII (notably Articles 42–44) of the United Nations charter.[18] The result was to breathe life into the stillborn collective security provisions and to raise the prospect that the U.N. might become the keeper of the peace in the new order, as its charter had suggested in 1945. That prospect received a boost in January 1992, when the heads of government of the Security Council states met in New York and endorsed the principle of collective security, and since then, through a series of U.N.–sponsored peacekeeping missions around the world.

There are three rejoinders that must be raised about the extension of Desert Storm and future collective security actions. First, there is the "Saddam factor." In addition to his generally outrageous behavior that virtually cemented world opinion against him, his invasion and conquest of Kuwait was the first time a member state of the United Nations had committed this illegal act against another U.N. member. In those circumstances, the U.N. could hardly do anything less than condemn and take action against an offender who evinced the observation that "a more compliant adversary may never be found."[19] Second, for the first time in the organization's history, the United States and the Soviet Union were not on opposite sides of the issue. The only potential opposing permanent Security Council member, China, abstained on the critical vote (Resolution 678) that authorized force. Third, the action was *not* a collective security operation as authorized under the charter. The charter calls for forces from the permanent membership of the Security Council, augmented as necessary by forces from other countries, to be permanently available to the organization. The idea, of course, is that awareness of these forces will serve as a deterrent to future aggressors. Such a force has been proposed; on any meaningful scale, it is unlikely to come about anytime soon.

This, of course, was not the model of Desert Storm. Rather, the U.N. was employed as the legitimating body for an ad hoc coalition formed under American leadership put together for the specific purpose of freeing Kuwait, after which it promptly dissolved. That method, authorizing force and calling for volunteers, is the method of the old League of Nations–style collective security. The problem, as the League experience showed, is that a potential aggressor has no idea what coalition will meet his hostile act, and is thus much less likely to be dissuaded in the first place. For Desert Storm to act as a collective security precedent, it will have more closely to approach the U.N. model. Moreover, this was really a U.S. operation where the coalition "was prized more for its political value than for any material strength" it provided.[20]

The third, and potentially most consequential, legacy occurred after Desert Storm ended. Following the defeat of Iraq, Iraqi Shi'ites in the south and Kurds in the north of Iraq rose in rebellion to overthrow Saddam Hussein, an action opposed by the U.S. government because it feared Iraq would be carved into three separate states that could not act as a counterbalance to Iran. The rebellions became a rout with genocidal potential, especially against the Kurds, who fled in large numbers to neighboring Iran and Turkey.

Cable News Network (CNN) coverage of the bleak plight of the Kurds galvanized the United States government to come to their aid. The result was Operation Provide Comfort, whereby the U.S. military occupied territory in northern Iraq, from which it excluded Iraqi forces, to allow Iraqi Kurds to return from their exile. (The Turks demanded their removal from Turkish soil; the Kurds would not leave without protection.) The only international authorization of this action was U.N. Security Council Resolution 688, which demanded that the Iraqi government allow "humanitarian organizations" to enter the Kurdish zones.[21] At that, the Soviet Union and China abstained because of the potential precedent established. In the summer of 1992, the precedent was extended to the Shi'ites in the form of a "no fly" rule enforced by Operation Southern Watch.

Operation Provide Comfort, for all its humanitarian intent, represented a violation of the territorial sovereignty of Iraq and asserted, at least implicitly, that there were limits to the sovereign ability of a government to deal with its own citizens. As such, it represented a clear obviation of the rights of states to exercise fully sovereign jurisdiction over their own territory, a basic international legal tenet since the Peace of Westphalia of 1648. Instead, it seemed to argue that there are boundaries around how states can suppress their minorities, an assertion of the rights of individual and groups against their governments, leaving the rights of states position as "the last refuge of a tyrant...that goes against the grain of our time."[22] The U.N.–sponsored action in Somalia (Operation Restore Hope), originally manned by U.S. troops is similar, in that no government invited the forces.

The United States has not argued that Provide Comfort or Restore Hope are precedents because of the enormous implications such precedents would represent. There are, after all, governments all over the world that suppress parts of their citizenry, and to argue the international system has the right or obligation to interfere in such actions with military force, a position espoused by U.N. Secretary–General Boutros Boutros–Ghali, would provide a very broad menu of future actions. The U.N. has, in a sense, embraced the concept by interceding in Cambodia, but that is only a peacekeeping exercise. It should not be surprising that the former Soviets and Chinese would not endorse this idea as principle: the Chinese would be a candidate for intervention in Tibet, and the former Soviet Union has the multiple potentials for ethnic violence, the situation in Nagorno–Karabakh only representing the tip of the iceberg. Whether Provide Comfort and Restore Hope ultimately represent an aberration or a "building block for the future" is a question that has great consequences for decisions about the future use of military force.[23]

Emerging Trends

Two other global trends are affecting the environment in which national security resides. The first is the result of the technological revolution, with profound effects on the structure and intertwining of world economies, the lethality and sophistication of military hardware and surrounding support, and the nature of global telecommunications. The second trend has to do with the growing importance of a number of transnational issues which have the effect of broadening conceptions of what constitutes national security.

The first set of factors is the result of the technological or Third Industrial Revolution: the quantum increase in the interrelated technologies of computing, telecommunications, and a series of derivative technologies.[24] Each of these interact: computing breakthroughs have translated into more rapid developments in telecommunication due to things such as digitization, miniaturization, and the like.[25] In turn, these advances have been evidenced in derivative areas such as new materials science and fiber optics which increase both computing and telecommunications capabilities.

The cutting edge in the production of high technology resides largely in the triumvirate of economic superpowers: North America, the European Community (EC), and Japan.[26] Indeed, much of the advantage that these first world areas have over the rest of the world is their ability to develop and commercially exploit emerging technologies, often by exporting manufacturing to near–first world states such as the Newly Industrial Countries (NICs) of the Pacific Rim.

There are several salient aspects of this phenomenon.[27] First, because science and technology are inherently international enterprises, the global spread of technology is aiding in the emergence of the truly international economic system. Global business has become truly global in terms of the ownership, management, labor force, and product composition of those companies doing business. The stateless corporation is indeed the wave of the future;[28] the recent furor over what is and is not an American automobile is just the most obvious manifestation of a trend which will in time cut across the productive spectrum.

This emerging international economy has some strong national security implications.[29] For one thing, globalized production means hardly any nation makes everything it needs, including all the components necessary for its weapons inventory. The United States' success with smart weapons in the Persian Gulf would have been difficult or impossible without Japanese microchips in the guidance systems. The first world is becoming so economically intertwined that it would be virtually impossible for any of us to fight one another (if we wanted to). Increasingly, we are all in this together to the arguable point that "global interdependence is rendering it difficult to define just what constitutes the national interest."[30] Talk about "economic warfare" is hyperbolic at best. Those outside the technological revolution are at a great disadvantage. Because technology fuels itself (this generation of computers creates the next one), if one is behind, the tendency is to fall further behind. As I have argued elsewhere, a prime reason that Mikhail Gorbachev felt the need to push reforms was the perceived need to knock down the barriers to participating in technology erected against an enemy.[31] At the same time, the countries of the third world will almost certainly fall progressively further behind in this competition, which they have scarcely joined at all.

This leads to the second impact of the technology revolution, which is the parallel revolution in military technology. As a conscious effort by the United States government, the tying of technology to war began during the Carter administration under the tutelage of Secretary of Defense (and former nuclear physicist) Harold Brown. Faced with the problem of confronting a numerically large Soviet opponent, he sought to negate Soviet quantitative advantage with American technological superiority, using technology as a force multiplier to offset Soviet numbers.[32] The result was the offset strategy put to great advantage in Desert Storm.

The offset idea employs all three aspects of the technological revolution. Advances in computerization and miniaturization allow the design of weapons that can be directed to targets with uncanny accuracy. Telecommunication advances in satellite reconnaissance and communication gave the United States a tremendous view of the battlefield while obstructing the opponents' view. At the same time, advances in materials science allowed the design a low radar visibility ("stealth") weapons platforms that the enemy could not see until they were upon him.

The effect of all this is twofold. On the one hand, the technological and scientific bases upon which this revolution exists is international and, in important ways, shared among the first world nations. Our interdependence means we cannot use these technologies to fight one another, but it does lead to the second observation: the gap between those who do and do not possess the most advanced technologies is growing.[33] As noted in the last section, the most obvious gap is between the military capabilities of first and third world countries.

The third technological area is specific to telecommunications and its impact on the way world events are viewed. More specifically, telecommunications have made the globe both smaller and more transparent. The most obvious manifestation is global television, the pioneer of which is CNN. CNN is not so much a global television service as it is an American television service operating globally, but already truly international television enterprises are in existence (the Independent Television Network or ITN) and more are on the boards.

I have discussed the impacts of television on national security elsewhere, but a key example may help explain the impact.[34] One obvious aspect of global television is to make events around the world public for all to see. This transparency applies most obviously to violent situations, including organized armed violence, especially when it involves government repression of groups within its own borders. In the past, it was possible to hide atrocities from the world and hence to deny their existence, but that is decreasingly the case. Governmental thuggery is now very difficult to conceal, and when it is revealed, the result is outrage and often a call for curative action.[35] For example, before CNN, it is unlikely that the world would have known of Saddam Hussein's repression of the Kurds; in that circumstance, there simply would not have been an Operation Provide Comfort, and the Kurds would have been left to fend for themselves. (The same is true of Bosnia's suffering first publicized by ITN.) That is, after all, the way the old system, based on the rights of states, operated. But can the principle of absolute state sovereignty survive the scrutiny of television? If the answer is negative, a whole new national security agenda may be opened.

The other general trend is the emergence of a series of transnational issues, some of which have obvious national security aspects and some of which require an expansion of what constitutes national security.[36] All share the commonality of being North–South areas of concern. In its attempt to redefine its role in a changing world, the U.S. Department of Defense has attempted to adopt a series of these issues to replace its more traditional NATO–WTO role. At the head of its list are three roles: counternarcotics, counterterror, and peacemaking and peacekeeping. All are nontraditional roles; none are roles for which the bulk of the United States military has any particular experience or expertise.

The counternarcotics effort has centered on the celebrated "war on drugs" initially announced by President Reagan and adopted and expanded by President Bush. The emphasis of this effort, which gives it a North–South flavor, is on the supply side of the drug problem, interrupting the supply of drugs entering the country by attempting to eradicate the crops from which drugs are refined and interdicting the finished products from entering the United States. The chief target has been the Andean nations of Colombia, Bolivia, and Peru, the sources of cocaine.

This role has been controversial on several grounds.[37] First, many experts believe it is misdirected, arguing that decreasing demand for drugs is the only viable long term solution. Currently, about 70 percent of the resources allocated for fighting the drug war goes to eradication and interdiction efforts, and the results have been less than impressive. Second, it is not clear that the military is the appropriate instrument for dealing with drug enforcement. At one level, the culture of coca plants (from which cocaine is refined) is a domestic social and economic matter in Peru and Bolivia, where virtually all coca is grown. Moreover, so-called narco–insurgencies exist to varying degrees in both producing countries and in Colombia, where the raw materials are manufactured into the finished product. These alliances between drug interests and revolutionaries create a difficult problem for the concerned governments and a nearly intractable situation for an outside party such as the United States.[38] These difficulties create a third source of controversy: the military itself is a less than an enthusiastic participant in the effort. When the idea of employing the military in the drug campaign was first raised, it was opposed by the Pentagon; only after it became clear that the military was in fact going to be used and that the drug wars were a source of continued funding in an atmosphere of constrained resources did the embrace occur.

Counterterror and peacekeeping–peacemaking operations are similar.[39] Countering terror is clearly a North–South problem akin to the drug problem. Just as most drugs are produced in the third world for consumption in the first world, the source of terrorism is basically in the third world (especially the Middle East). Terrorism is not a large problem for the United States, because essentially the only Americans potentially subject to terrorist actions, prior to the attack on New York's World Trade Towers, are those who reside in third world areas where terrorism occurs. The American government has small elite units, such as Delta Force, who are specially trained for this role. Separating warring parties and then supervising their cessation of hostilities, peacemaking and peacekeeping, is a growing enterprise in the emerging world order. As the political map is redrawn, especially in the old second and third worlds, violent friction is increasing among ethnic and national groups formerly restrained by authoritarian governments. The violence in what is left of Yugoslavia and in parts of the former Soviet Union such as Nagorno–Karabakh are the most visible manifestations of this problem; there will doubtless be more in the future as democracy and national self–determination spread more generally. To this point, the United Nations appears to be the focal point for international efforts in this area, a role consciously embraced by Boutros–Ghali.[40]

In addition to these national security problems with at least arguable military components, a number of other transnational issues have entered into the national security debate. What distinguishes these issues is their nonmilitary character: they do not resemble traditional vital interests in the sense that they are problems over which the nation is or should be willing to go to war. Rather, they are problems that cannot be solved independently by nation–states and which, if they are not solved, pose threats to the quality of existence within states.

Environmental degradation is the most dramatic case in point, and it is not coincidental that a major international congress on the problem met in the summer of 1992 in Rio de Janeiro. Degradation of the environment is a truly international problem; pollution of the ecosystem is global and beyond national solution. Moreover, it is a North–South problem; the balance of pollution is moving from the first world, where fossil–fuel consumption has been a major contributor to pollution, toward the third world, where such things as the emergence of smokestack industries and the denuding of the rain forests contribute to the problem.

Consequences of Change

All of these changes have major impacts on the way that we view national security and the national security problem. The collapse of the Cold War renders the old constructs built to deal with the East–West confrontation dubiously relevant and the strategies and forces developed for those problems vulnerable to criticism. At the same time, the contours of the emerging system remain murky, as concrete, durable replacements for the old system of threats continue to be elusive.

From a strategic point of view, the most obvious impact has been to negate the upper end of the conflict spectrum: strategic nuclear war and a large conventional, possibly nuclear, war in Europe. The reasons for this are clear enough and have already been discussed in the last section. The consequences are contentious. Does the end of the Cold War validate the strategies employed by NATO over the past forty years? Because the purpose was deterrence and war did not occur, there is an initial inclination to declare victory for the strategy. The perversity of deterrence, of course, is that one cannot demonstrate its effectiveness without committing the logical fallacy of affirming the consequent; the most one can say is that the strategies and peace coincided. If one makes the assumption that Western strategy toward the Soviets had something to do with the absence of nuclear or conventional war with Soviet Union, does the strategy have any value now that the Soviet opponent no longer exists? Here one must disaggregate the two aspects of strategy, nuclear and conventional, to render a judgment.

The nuclear strategy of deterrence based on mutual societal vulnerability and retaliatory threats probably maintains some salience. Russia, Belorus, the Ukraine, and Kazakhstan do possess the strategic arsenal formerly held by the Soviet Union, and although it is difficult to see why any of them would attack the United States with those weapons, the deterrent effect of our own weapons may provide some comfort during a transition period in the former Soviet Union.[41] At the same time, the specter of horizontal proliferation remains near the horizon, and the concepts that underlay Soviet–American deterrence may be transferable, possibly in altered ways, to deterring third world proliferators. A number of analysts have argued that nuclear deterrence Cold War–style occurred within a specific Western cultural heritage that does not apply in the third world, but that is only a hypothesis.[42]

If the bases of nuclear strategy continue to have some value, the shrinking threat does render questionable the size of arsenals on both sides. If Russia no longer considers the United States the enemy and hence does not target us, then at what does it aim the 7–8,000 warheads that remain under the Strategic Arms Reduction Talks (START) agreement? Similarly, it is hard to defend the details of an American targeting plan (the Single Integrated Operational Plan, or SIOP) that identifies 60,000 targets for potential destruction among the successor states to the Soviet Union.

The retention of nuclear forces by the United States and Russia notably larger than those of anyone else probably does make strategic sense. Nuclear weapons have been the membership dues for great power status in the second half of this century. Nuclear weapons are about all that is left of claims by the former Soviet Union to great power status, and it would do little good to deprive them of that claim to status. At the same time, a number of the states that may acquire nuclear weapons are near former Soviet territory, making deterrence a more lively concern for the Russians than the Americans. As the world's remaining global military power, the United States will retain a robust arsenal for deterrent purposes. The familiar "how much is enough?" will enliven the debate over arsenal size.

The conventional side of the ledger is more problematical. In one sense, the application of power as envisioned in Europe in the Persian

Gulf War appears to vindicate the strategies and forces prepared for central Europe. Because of this, for instance, the Army is resisting "lightening" its force mix (moving away from heavy armor and artillery to light, highly mobile infantry), and it believes the active–reserve component mix two–to–one should be retained. The other side of the coin is that large, heavy forces are simply not as necessary as they once were. As the former Soviet forces shrink in size and retreat to their constituent republics, the need for large forces deployed forward is rendered questionable. There are a few large, heavily armed armies in the world (India, for instance), but it is hard to imagine why the United States would engage such forces. It may be that "changes in the character of relevant military power" will require adjustments to the force mix.[43] Moreover, if a legacy of the Persian Gulf War is indeed the overwhelming qualitative superiority of American forces, size will not be so important, as long as the offsetting technological advantage is maintained.

In a general sense, the shrinkage in the threat and its force consequences is generally accepted, even within the military itself. The transition is still tortured, and its implications will be resisted, for two reasons. First, although there is a visceral acceptance of the need for change, it quickly becomes very personal within individual military specialties that are heavily represented in command structures. Infantry and armor generals, in other words, chafe when they are told the new world order does not require many soldiers or tanks. Second, uncertainty about the kinds of threats that will be encountered counsels caution. Although no one can readily identify plausible short–term scenarios where the strategies and forces designed for NATO would apply, that does not mean they will not appear. If they do and the capabilities have been dismantled, the result could be disadvantageous. This of course, is the argument currently being put forward by the Defense Department to justify minimizing cuts; while there are elements of a smoke screen behind which to hide foot dragging in this, there is also enough of an element of possibility to moderate change.

The final part of the spectrum is the lower end, contingencies mostly in the third world involving forces generally of a lesser nature that those involved in a NATO Europe battle and often conducted by "light" assault forces, special operations forces, and often supported by sophisticated land– and sea–based aviation and naval assets. This range of contingencies does not closely resemble the more familiar European scenarios in terms of interest or threat definition, strategic approach, or supporting force composition and mix.

The most obvious difference between European and third world scenarios is that, generally speaking, American vital interests are not clearly or unambiguously affected. The apparent exception, of course, is where petroleum is present in large quantities. Nonetheless, when either internal or international instabilities emerge, it is not always clear whether the United States has sufficient interests to become militarily involved in places such as India, Sri Lanka, or Ethiopia, to borrow from one list.[44] The absence of clearly defined vital interests means that these situations have not been ones for which the United States extensively prepared, either conceptually in terms of foreseeing them or in preparing for them. This is at least one of the reasons why uncertainty is such a prevalent part of the strategic landscape; we expect to be surprised and thus have to react flexibly to unforeseen (some in the government would argue unforeseeable) contingencies for which we must have maximally flexible strategies and forces.

Familiar strategies designed for strategic nuclear deterrence or NATO do not obviously apply in the third world. The transferability of deterrence ideas to the third world is an open question; the transfer of NATO strategy to the third world was thoroughly debunked in Vietnam. In a few instances such as Desert Storm, there may be some transferability, generally where the opponent was trained and equipped and thus resembles Western forces. Where he does not, as in Southeast Asia and Afghanistan, the strategies need to be modified.

The force mix that supports the lower end of the spectrum is also quite different. In most instances of regional conflict or internal conflict, the United States is unlikely to want to become heavily involved with American ground forces. The American people have a limited appetite for the massive use of American soldiers, especially when overwhelming American interest are not clearly involved. The case for compelling interest cannot often be made in the third world; Desert Storm was the exception. In the case of involvement in internal conflicts, large American forces are inappropriate and may even be counterproductive.

This suggests that the forces needed to support the lower end of the spectrum will be of four types. First, we will need special operations forces for insertion into volatile situations (e.g., the Liberian rescue of 1991) or in support of counterinsurgency efforts (e.g., El Salvador). Second, we will require highly mobile, rapidly deployable airborne, ranger, and marine contingents for quick insertion into relatively short contingency operations (e.g., Operation Just Cause in Panama). Third, we will need highly mobile and capable air and sea forces. These will be useful for assisting others by establishing air or sea superiority, for rapidly deploying our own or U.N. peacemaking or peacekeeping forces into third world contingencies, or simply for shows of force. Fourth, we will need units trained in military police–like tactics for use in future actions like Somalia. Other roles and missions will have to be otherwise justified.

Conclusions: The Emerging System

That the world is changing rapidly and that the threats to American national security are shifting is no longer a noteworthy observation, nor is the relative permanence of American interests in the world. We are coming to understand the process of change that has occurred and how it has had an impact on the old ways of looking at the world and organizing our responses to it. What is not so clear is what the new system will look like.

The answers are not yet clear, but the questions may be. I would suggest that something of a road map for viewing the future can be constructed by asking five questions about the evolving system. Two of these deal primarily with how the United States chooses to face the world. Three deal with how the system, influenced by the United States, will shape the future.

The first American question is: *Will the United States approach the system from an activist, expansionist, or a neo–isolationist perspective?* Having won the Cold War, the host of domestic priorities seem all the more compelling, and there is in the air a thread of opinion that suggests a diminished role for the United States and an expanded role for others. Counterbalancing this trend is the realization that as the remaining global power, the U.S. has an obligation to lead and the parallel realization that the world is so intertwined economically, and otherwise, that isolation is no longer possible.[45] Because the United States is such an important nation, however, the degree of its participation is a major variable in how the system evolves.

A second primarily American determination is: *How will the interest–threat mismatch be resolved?* This is a question the answer to which is framed especially in national security terms, because it tautologically raises the question of where and under what circumstances the United States will be willing to employ armed force to quell violence and instability in the world. As indicated already, the new system is unlikely to suffer a shortage of violence and instability; the problem is that most of that violence will occur outside those areas historically deemed vital. The emerging system, led by the United States, may choose to ignore problems in marginal areas as we have done in the past (e.g., Ethiopia, the Sudan). At the same time, our collective guilt may force us to reconsider where and when to put ourselves in harm's way (e.g., Cambodia).

This leads to the third question, which is largely systemic: *Will the apparent trend toward collective security continue, or will we revert to some form of collective defense?* From the vantage point of early 1992, collective security is clearly on the upswing, but three questions about its role remain in limbo.[46] The first is what kind of collective security system will there be? The U.N. charter prescribes a very formal system of great power dominance that has never been implemented; the alternative is less formal, more ad hoc arrangements of the Desert Storm variety. The second question is what kinds of collective security actions will there be? In the past, the dominant form has been peacekeeping, and the question is whether that will expand to peacemaking and even beyond to large–scale deployments. The third question is against whom will collective security be directed? Desert Storm's coalition to stop aggression offers one category, and invited peacekeeping as in Yugoslavia and Cambodia represents another. There may be even more active uses both within and among states in the future. The fourth problem is who will pay for this activity. By the end of 1992, for instance, the U.N.'s resources were clearly overextended and its capability exceeded unless additional human and monetary resources are volunteered.

This leads to the fourth question: *Will the new world order be marked by a greater assertion of the rights of individuals and groups against states, or will it revert to the old Westphalian order where the rights of states are supreme?* This is the question raised by Operation Provide Comfort; as stated at the time, it opens a whole new range of situations in which the international order, including the United States, might be called upon to come to the aid of repressed segments of populations. The problem is that such situations are often long–term, bitter, and tenacious problems for which there are no easy solutions. Including such situations in the new order's menu potentially open a Pandora's box of unforeseen horrors that my help explain why the term "new world order" has largely slipped from official usage. The ambivalent, hesitant international response, for well over a year, to the horrors of Bosnia and Herzegovina indicates how difficult the problem is.

The fifth and final question, however, is one that bedevils attempts to revert to the old order: *How will the system deal with an effectively smaller and much more transparent world?* Ignoring violence and atrocity in the third world used to be a relatively easy thing to do; it occurred but was scarcely ever within the public conscience. How many Americans, for instance, knew of the mutual slaughter of Watutsi and Bahutu tribesmen in former Rwanda–Burundi until those countries were separated in 1968? Global television, still in its infancy, has changed all that. Once again, Operation Provide Comfort provides a precedent of sorts. The reason the world knew of the plight of the Kurds was because it was very publicly broadcast across the world compliments of CNN. Without that coverage, the world quite probably would have ignored the Kurds' fate; the communal slaughter over Nagorno–Karabakh would hardly have been known had it occurred ten years ago. The point, of course, is that world problems will increasingly be known in the emerging system; that system's capacity to ignore man's inhumanity will be all the more difficult to ignore as well.

Notes

1. See Donald M. Snow, *National Security: Enduring Problems in a Changing Defense Environment*, 2nd edition (New York: St. Martin's Press, 1991), 4–6 for a discussion.
2. For an overview, see Daniel S. Papp, *Soviet Policies Toward the Developing World During the 1980s: The Dilemmas of Power and Presence* (Montgomery, Ala.: Air University Press, 1985), 1–26.
3. Two recent articles make this point: Zbigniew Brzezinski, "Order, Disorder, and U.S. Leadership," *Washington Quarterly* 15 (Spring 1992): 5–14 and David Abshire, "Strategic Challenges: Contingencies, Force Structures, Deterrence," *Washington Quarterly* 15 (Spring 1992): 33–42.
4. For an overview of the issues, see Joseph S. Nye, Jr., "What New World Order?" *Foreign Affairs* 71 (Spring 1992): 83–96.
5. For a recent assessment of the prospects for Russia by a Russian, see Andrei Kozyrev, "Russia: A Chance for Survival," *Foreign Affairs* 71 (Spring 1992): 1–16.
6. Todd R. Greentree, *The United States and the Politics of Conflict in the Developing World* (Washington, D.C.: U.S. State Department Center for the Study of Foreign Affairs, 1990), 13.
7. Richard N. Haass, "Regional Order in the 1990s: The Challenge of the Middle East," *Washington Quarterly* 14 (Winter 1991): 182.
8. Colin S. Powell, Statement of the Chairman of the Joint Chiefs of Staff before the Sub–Committee on Defense Appropriations, Committee on Appropriations, U.S. House of Representatives, February 19, 1991.
9. Mark N. Katz, "Beyond the Reagan Doctrine: Reassessing U.S. Policy toward Regional Conflicts," *Washington Quarterly* 14 (Winter 1991): 174.
10. See Donald M. Snow, *The Shape of the Future: The Post–Cold War World* (Armonk, N.Y.: M. E. Sharpe, 1991), 113–131.

11. For contrasting views of this process, see Andrew Nagorski, "The Intellectual Roots of Eastern Europe's Upheaval," *SAIS Review* 10 (Summer/Fall 1990): 89–100 and Sergei Karagonov, "The Year of Europe: A Soviet View," *Survival* 32 (March/April 1990): 121–128.

12. For an assessment from the former Soviet Union, see Sergei Rogov, "International Security and the Collapse of the Soviet Union," *Washington Quarterly* 15 (Spring 1992): 15–28.

13. Bobby R. Inman, Joseph S. Nye, Jr., William J. Perry, and Roger K. Smith, "Lessons of the Gulf War," *Washington Quarterly* 15 (Winter 1991): 68.

14. William J. Taylor and James Blackwell, "The Ground War in the Gulf," *Survival* 33 (March/April 1991): 240.

15. Ibid., 239.

16. Inman et al., "Lessons of the Gulf War," 62.

17. Strobe Talbott, "Post–Victory Blues," *Foreign Affairs* 71 (Winter 1991–92): 60.

18. For the text of the relevant sections, see Snow, *Shape of the Future*, 213–218.

19. Inman et al., "Lessons of the Gulf War," 57.

20. Michael Brenner, "The Coalition: A Gulf Post–Mortem," *International Affairs* 67 (October 1991): 670.

21. Paul Leavis, "UN Votes to Condemn Handling of Iraq Rebels," *New York Times*, April 6, 1991, 6 and "Survival Is Harsh, Recovery Slow in Hard–Hit Areas," *UN Chronicle* 28, (September 1991): 16.

22. David A. Korn, "Iraq's Kurds: Why Two Million Fled," *Foreign Service Journal* 68 (July 1991): 24.

23. Lawrence Freedman, "The Gulf War and the New World Order," *Survival* 33 (May/June 1991), 24.

24. Snow, *Shape the Future*, 58–64.

25. For a good review see Tom Forester, *High–Tech Society: The Story of the Information Technology Revolution* (Oxford: Basil Blackwell, 1987).

26. Daniel F. Burton, Jr., Victor Gotbaum, and Felix G. Rohatyn, eds., *Vision for the 1990s: Strategy and the Global Economy* (Cambridge, Mass.: Ballinger, 1990) provides a good overview.

27. For a good literate review on competitiveness, see Debra L. Miller, "A Domestic Agenda to Strengthen America," *Washington Quarterly* 15 (Spring 1992): 205–224.

28. See William G. Holstein, "The Stateless Corporation," *Business Week* (May 14, 1990): 98–105.

29. Donald M. Snow, "High Technology and National Security: A Preliminary Assessment," *Armed Forces and Society* 17 (Winter 1991): 243–258.

30. John P. Cregan, "Building an American Consensus: A National Interest Trade Policy," *Vital Speeches of the Day* 56 (June 1, 1990): 511–512.

31. Donald M. Snow, "Soviet Reform and the Technological Imperative," *Parameters* 20 (March 1990): 76–87.

32. For a description, see William J. Perry, "Desert Storm and Deterrence," *Foreign Affairs* 70 (Fall 1991): 69.

33. The security implications of the phenomenon are discussed in Theodore H. Moran, "International Economics and National Security," *Foreign Affairs* 69 (Winter 1990–91): 90.

34. Snow, *Shape of the Future*, 64–70.

35. See Herbert S. Dordick, "New Communications Technology and Media Power," in Andrew Arno and Winral Dissanayake, eds., *The News Media in National and International Conflict* (Boulder, Colo.: Westview Press, 1984), 38.

36. Robert D. Hormats, "The Economic Consequences of the Peace, 1989," *Survival* 31 (November/December 1989): 487.

37. For a critical summary, see Peter R. Andreas, Eva C. Bertram, Morris J. Blachman, and Kenneth E. Sharpe, "Dead–End Drug Wars," *Foreign Policy* (Winter 1991–92): 106–128.

38. See Scott D. McDonald, *Mountain High, White Avalanche: Cocaine and Power in the Andean States and Panama*, CSIS Washington Papers no. 137 (New York: Praeger, 1989).

39. See Donald M. Snow, *Distant Thunder: Third World Conflict and the New International Order* (New York: St. Martin's Press, 1992), chap. 5.

40. For a discussion of the implications of Operation Provide Comfort, see Edward C. Luck and Toby Trister Gati, "Whose Collective Security?" *Washington Quarterly* 15, (Spring 1992): 43–56. Boutros Boutros–Ghali has articulated his position in at least two major places, *An Agenda for Peace: Preventive Diplomacy, Peacemaking, and Peacekeeping* (New York: United Nations, 1992) and "Empowering the United Nations," *Foreign Affairs* 71 (Winter 1992–1993): 89–102.

41. Kozyrev, "Russia: A Chance for Survival," 14–15.

42. Snow, *Distant Thunder* discusses this in chap. 6.

43. William Pfaff, "Redefining World Power," *Foreign Affairs* 70 (1990–91): 46.

44. Yezid Sayigh, *Confronting the 1990s: Security in the Developing Countries*, Adelphi Papers no. 251 (London: International Institute for Strategic Studies, 1990), 10.

45. For an overview of the isolationism–internationalism debate, see David C. Hendrickson, "The Renovation of American Foreign Policy," *Foreign Affairs* 71 (Spring 1992): 48–63.

46. Richard Rosecrance, "A New Concert of Powers," *Foreign Affairs* 71 (Spring 1992): 64–82.

5

A Farewell to Arms? The Military and the Nation-State in a Changing World

Christopher Dandeker

In this chapter, I explore two closely related themes. One focuses on the prospects for the decline in the role of force in international relations. The immediate context for this theme is the end of the Cold War, the death of communism, and the apparent triumph of liberal capitalism, together with a corresponding growth of a larger core of industrial capitalist states which are less inclined to war. It is possible that this community of states will be more able than previous ones to operate in a context of collegiality organized through the United Nations (U.N.) and, in the future, through a U.N. Security Council that is likely to be reconstructed to take into account the economic and geopolitical realities of the twenty–first century rather than those of 1945.

The second theme deals with the implications of these shifts in international relations *and* with changes in the social structure of advanced industrial societies as these affect military establishments. I am concerned, in particular, with the prospects of developing smaller more flexible armed forces equipped, and tasked, to deal with a far broader range of security risks than the armed forces of the Cold War (or, for that matter, of any earlier historical) period.

The "Warless Society" and the Development of Modern Sociological Theory

It is possible, but misleading to examine the first theme solely in terms of current events, however dramatic these may be. We are better served, I think, to consider the links between modernity and war against the background of the historical development of sociological theory. The idea of a warless or postmilitary society, produced by a decline in the threat of war between modern industrial states, played a key part in the classic writings of the nineteenth–century founders of sociological thought. Indeed, most sociologists since then have paid insufficient attention to war and military power precisely because they accepted some variant of the "warless society" paradigms as developed in either the Marxist or liberal theories of industrial society.[1]

Both traditions are equipped with reductionist views of warfare. They collapse military power to other elements of modernity. In the case of Marxism, the element in question is capital accumulation and the dynamics of class conflict. Marxist sociology regards warfare and military organization as aspects of the development of political struggle in class societies. Far from being a constitutive feature of a world of competing states—or, in the modern age, of nation–states—war is contingent upon the existence of class divisions based on property relations. The socialization of the means of production and the abolition of class divisions, preferably on a world scale, would remove the socioeconomic basis of war and military organization and preface the establishment of a peaceful confederation of socialist states. This view constitutes a mirror image of the liberal vision found in the theory of industrial society; for one capitalism is the root of war, for the other it is the root of peace. Recently, with the fading of the intellectual and political power of Marxism, it is liberal social theory that has provided the most popular and persuasive account of the supposed decline of war and the prospect of what Martin Shaw has called a postmilitary society.[2]

Why should the system of liberal capitalism foster peace rather than war? In a critical discussion of the question, John Mearsheimer has pointed to three possible mechanisms. First, such a system makes its participants prosperous, and thus less likely to turn to war to resolve their economic difficulties. Indeed their very prosperity gives them more to lose by turning to war. They have a vested interest in peace, which promotes international cooperation, further prosperity, and more cooperation in a self–reinforcing feedback cycle. Second, economic interdependence promotes mutual vulnerability and this means that states acquire interests in cooperation and in devising trust and confidence building measures. Third, although it is accepted that not all economic liberals argue this, some do follow the original Comtean vision of a united humanity under a world state. Tying all these mechanisms together is the theoretical premise that the root of conflict is economic scarcity and that the most important motive of individuals, including the state as a political actor, is the resolution of this problem.[3]

It is, therefore, rather ironic that the mini–renaissance of the sociology of war and military power, which can be dated from the mid–1980s, arrived just at a time when, at least as far as the capitalist core countries are concerned, and notwithstanding the Gulf War, the utility of force as a means of regulating international relations has come under scrutiny and doubt.[4] In short, sociologists' discovery of the realist approach to international relations has occurred belatedly and at a time when the agenda has changed. War and military power focused on the modern nation–state are supposed to be of declining relevance in the increasingly "globalized" interdependent planet which we inhabit. So, with the death of communism and the emergent hegemony of the values of market capitalism and liberal democracy, it has been argued by Fukuyama that we are witnessing perhaps "the end of history" with all that promises in the way of a pervasive and enduring peace.[5]

This is the broader theoretical context in which one should place Charles Moskos's recent formulation of the "warless society" thesis, that "war—at least between superpowers and major European powers—is no longer the principal, much less inevitable mode of conflict resolution."[6] While the end of the Cold War has given his thesis a particular resonance today, it has deep roots in the history of sociological reflection on the relation between modern society and war. Unlike Fukuyama, who has been accused of Hegelian complacency, Moskos warns that we should not expect the end of war across the planet. War and military power will be extruded from the relations between the advanced capitalist societies. But they will instead become focused on two dimensions of the interface between core and peripheral countries, on the relations between the developed capitalist world and the underdeveloped nations *and* amongst the less developed peripheral countries engaged in regional disputes. At the same time, however, the relatively peaceful community of the advanced industrial societies is likely to grow and take in more of the planet as we head into the twenty–first century (although, as I shall argue later, there is a case to be made for the persistence of ethnic, regional, and separatist violence with implications for armed forces in core nation–states). Thus his position is optimistic, but not utopian.

It can be argued that the prospects for the spread of liberal capitalism from developed to developing nations look brighter than they have in the past. Yet the use of force against regional dictators who stand in the way of the triumph of liberal values will continue to be necessary. Most likely, this can take place only under the aegis of an international community no longer riven by the bipolar superpower conflict, and indeed guaranteed militarily (if not economically) by the only superpower, the United States. The developed powers themselves have little reason now, and will have even less reason in the future, to resort to military means to resolve disputes with those who are economically and politically rather like themselves. Winners and losers there will continue to be. But who wins and who loses will be determined by "soft" social and economic power mechanisms rather than "hard" military ones, and their operation will increasingly be constrained and mediated by supranational organizations. The 200-year-old reign of the sovereign nation-state is drawing to a close. With some qualification then, it could be argued that the claims of liberal social theorists in the nineteenth century were right. They were just stated a little prematurely.

Mass Armed Forces in Decline: The Social Organization of Military Power of Western Nation–States since 1945

It is in the light of these larger social theories that we need to assess the impact of the end of the Cold War (and a wider set of global social processes) on the military establishments of the core industrial capitalist countries. The key question is whether a warless society entails a postmilitary era or rather, more simply, just a society equipped with a different type of armed force from that characteristic of the Cold War era. To answer it, we must look back to long–term trends in the relations between war, military power, and the nation–state. Doing so allows us to identify three phases in the social organization of military power: one stretching from the French Revolution to the Second World War—the age of the mass armed force; the second, from 1945 to 1989—the age of the force in being—rooted in the Cold War and the strategic revolution focused on the invention and development of nuclear weapons; and the third, beginning with the end of the Cold War, encompassing broader global changes, yet recognizing the persistence of nuclear weapons and the continuing (if diminished) threat of global war—an age of smaller, more flexible forces organized along the lines of a cadre–reserve system.[7] It is this third phase that I will want to address in the greatest detail beginning in the next section by focusing on the global changes which have brought it about and afterwards by considering the implications of these changes for military organization in advanced industrial societies. Yet, to begin, let me sketch in the first two of these phases, drawing on the work of Morris Janowitz, Jacques Van Doorn, Gywn Harries–Jenkins, Philipe Manigart, and most recently, of Charles Moskos, Bernard Boëne, and James Burk.

In Western nation–states since 1945 a substantial restructuring of military power has occurred—what military sociologists in the 1970s such as Janowitz and Van Doorn referred to as the decline of mass armed forces and the emergence of a force in being.[8] As Burk has argued in an excellent analysis of this question, these two types of military organization can be contrasted in terms of three dimensions: mission, force structure, and citizen service.[9] The primary goal of the mass armed force is to achieve military victory wherever it is deployed. The mission of the force in being is not so straightforward. As Burk suggests, drawing on Janowitz's classic analysis of the constabulary force, while no armed force deliberately seeks to fail in achieving its objectives, in this case the main goal is to deter the outbreak of international conflict in the first place and to limit its scope should conflict occur. Indecisive outcomes are acceptable if that is the price to be paid for a political settlement.[10]

In terms of force structure, the mission of war winning leads mass armed forces to build numerically large forces in wartime as the relatively simple division of labor dependent on rifle infantrymen—paralleling the mass production work force in the civilian economy—allows for a reliance on nonprofessionals conscripted in time of war and demobilized in time of peace. A relatively small, professional cadre remains on active duty for war preparation and to train conscripts.

In contrast, for the force in being, the mission of peacekeeping requires it to be permanently mobilized while its dependence on technologically sophisticated weapons of mass destruction means that the military division of labor is much more complex. Consequently, greater emphasis is placed on longer service professionals instead of short–term conscripts. Even if the formal system of conscription is retained, it is diluted in the direction of professionalism. (Thus, Burk shows that in Europe over the twenty-five years from 1961 to 1986 the average number of months which conscripts have to serve decreased from a minimum of eighteen months to a minimum of twelve.[11]) The size of the force in being is moderate, and while larger than the mass armed force in time of peace it is smaller than the mass armed force at war. Because it is permanently mobilized, there are relatively small fluctuations in the size of the force in being over time.

There have been three main causes of this broad shift in the social structure of military power from the mass armed force to the force in being. First, modern weapons technology require skill levels which limit the usefulness of short–term conscripts. Furthermore, in the nuclear age, the destructive power of modern arsenals, in addition to their expense, means that it is difficult to retain war winning as the defining mission of armed forces. Deterrence places constraints upon the goals of military organizations.[12]

Second, since 1945 shifts in international relations, and particularly the collapse of the colonial empires of Britain, France, and Belgium altered the military requirements of these states' armed forces. The growth of third world nationalism and the establishment of numerous new nation–states limited the utility of force in serving the political aims of ex–colonial powers, while the Cold War focused military efforts on the development of a force in being to face the Soviet threat.

The third cause of the decline of mass armed force is sociocultural change. Specifically, growing affluence and higher levels of education in the post–Second World War era promoted popular reluctance to forego the benefits of consumerism for military service, a more individualistic climate and changing (less accepting) attitudes towards authority, and a reluctance to express a commitment to the national state through military service.[13]

The third trend in particular can be seen as undermining the basis for any military establishment. Nevertheless, the Cold War provided a rationale and political will to maintain a relatively large professional military establishment. The need for effective deterrence in a nuclear world organized around bipolar conflict clearly justified the existence of the force in being. With the end of the Cold War, this justification is called into doubt and heralds further changes in the social organization of military power which I shall examine presently. Yet it is essential before doing so that we place this geopolitical development in the broader context of other equally fundamental social changes in the "new times" of our modern era.

Globalization and the New Times of Late Modernity

It will be recalled that many sociological theorists have long believed that a "warless society" was the natural outcome of the historical development of advanced capitalist society. While the end of the Cold War has triggered a revival of the thesis, I want to suggest that this geopolitical event is merely one, albeit important, feature of a much broader and far–reaching social transformation that has called the role of armed forces into question (and more besides). It is tempting to suggest that the transformation is leading to the establishment of a postmodern world society. Yet I think, following Anthony Giddens, that late or high modernity is a more apt and accurate term. What we are experiencing is an accentuation of past trends and not as sharp a break with them as the term postmodern would suggest.[14] In any case, we may speak of these as new times, and we may indicate a number of related social processes which define what these new times are about.[15]

The first of these is "globalization," by which I mean the growth of an interdependent world economic system, coordinated through transnational corporations and a range of other international and transnational organizations, the operations of which have been facilitated by the electronic revolution in the means of communication. This process has spread the Western systems of economic capitalism, political democracy, technology, and military power to other parts of the globe. It has at the same time brought about a relative decline of Western hegemony in the Pacific. And it has caused an "intensification of worldwide social relations which link distant localities in such a way that local happenings are shaped by events occurring many miles away and vice versa."[16]

Second, and connected with globalization, are processes of regionalization and transregionalization. While regional differentiation between core and peripheral areas is hardly new in economic history, what is novel is that overlapping regional specializations now develop along quite different dimensions. Complex hierarchies of regional specializations develop in which relations of economic superiority can be quite out of kilter with relations of military superiority. At the same time, the global economy means that strategies of autarky are ruled out, while transregional coordination is ruled in, for those responsible for managing economic development.

Third, in the context of the global world system, the number and types of international actors are increasing. Sovereign states are now joined by transnational corporations (although these do not exercise sovereign power as do states) and other international and transnational actors equipped with the means of global reach, including emergent politico–economic blocs, notably, the "big three" of North America, Japan, and the European Community. The full significance of this trend is difficult to assess. On the one hand, with the end of the Cold War and the bipolar world of competing blocs, the world has become multipolar, creating a geopolitical system that is more unpredictable and less stable, at least in terms of threat assessment, as opportunities for independent action by nation–states increase. (This increase may perhaps be offset by better surveillance or monitoring capacities developed by the more advanced industrial states.) On the other hand, for many states, global economic pressures will force them to become attached to one or another of the big three economic blocs or, in the longer term, to create other blocs. The crucial and paradoxical issue arising here is the extent to which processes of bloc formation rooted, in part, in attempts to take economic and political advantage of the processes of globalization might lead to mercantilist–like processes of "de–globalization," a point to which I shall return.

Fourth, one can point to the increasing impact of international law on the conduct of states. This can be observed most immediately perhaps in the European context when one examines the role of the European Community and Court of Human Rights. But there are broader trends at work, more relevant to the use of armed force in the post–Cold War context. There seems, for instance, to be a wider questioning of the legitimacy of the unilateral use of military force as a means of resolving international disputes and, as the case of the Iraqis' treatment of the Kurds illustrates, there is also a greater expression of doubt about the sanctity of the sovereign right of states to govern their internal affairs. There is certainly reluctance within the international community to accept that states may treat their own people living within their territory as targets in a "free–fire zone." This questioning has led the democratic core of advanced capitalist countries to consider mounting more robust interventions in favor of democracy (and human rights) around the globe. The question is what criteria might emerge in order to guide their undertaking such actions, given that significant resource and legitimacy constraints work against any "blank check" interventionism. As the *Economist* pointed out recently, three criteria—which move away from the 1973 U.N. rulings on these matters—could be as follows: first, that there is an unambiguous demand that a people or its representatives want such an external intervention; second, that such an intervention is likely to be successful; and third, that the intervention is in the interests of the intervening power(s).[17] Underlying these criteria, however, is the fundamental problem of national sovereignty: Should the international community be prepared to intervene in disputes when the parties do not consent to, or indeed reject, such an initiative? Specifically, should such interventions override national sovereignty when individual or minority rights are threatened to such a degree as to challenge norms of "civilized" behavior? Quite possibly, moves in this direction will encounter the charge of "neo–imperialism." Still, one could argue in defense against the charge that the development of U.N. security mechanisms are wholly appropriate in an age of globalization. In any case, these kinds of interventions are made more likely by the end of the Cold War and the erosion of the bipolar limits on past U.N. actions. Thus in the four decades leading up to 1988, the U.N. endorsed only thirteen peacekeeping operations, whereas since that time it has authorized no fewer than fourteen more. In 1992, the number of troops involved in U.N. operations rose from 12,000 to 50,000.[18] Already, after the Gulf War at least four such interventions can be identified, in Iraq, on behalf of the Kurds, Ethiopia, Zaire, and in what was Yugoslavia.

Fifth, as something of a counterreaction to the growth of a global world system and the impact of transnational and international organizations, one can observe the development of nationalist and regional separatist movements with demands for political autonomy and independence. The significance of these developments is that interstate wars in these new times are *not* likely to be the most typical form of collective political violence; rather these will be supplanted by what Moskos has termed subnational violence. Indeed, there are those who suggest that subnational violence will not be confined to peripheral states but is destined to become a significant feature even within core nation–states. Thus Britain's security problems stemming from the conflict in Northern Ireland will cease to be atypical. Martin Van Creveld has gone as far as to argue that subnational and ethnic movements will lead to the decline of the state itself and of the state's armed forces.[19] In any case, I shall argue that the implications of these developments for the nature and function of armed forces in the advanced societies will be profound.

These movements within nation–states can be viewed as nationalisms of frustration against these larger political and economic processes. This has led some observers to view regionalism and political integration at the inter– and especially the transnational level as the erosion of the nation–state from above and below.[20] However, for the underprivileged peoples of the world, maintaining their status as independent nation–states continues to provide the principal means for extracting the most favorable deal from the privileged societies, international and transnational organizations that control the bulk of the world's resources. Thus, how one views the much vaunted decline of the nation–state may depend upon one's vantage point, looking out from the big three blocs or looking hungrily into them from the less economically privileged parts of the globe.

What are the consequences of these processes for the modern, developed nation–state? Let me point to two. First, as David Held suggested recently, it can be argued that relations of economic, political, and cultural interdependence across the globe are undermining the sovereignty (the legal/constitutional independence to make decisions) as well as the autonomy (the effective power to implement decisions) of nation–states in regard to all aspects of their security: economic, social, politico–military, and ecological.[21] Of course, relations of global interdependence have been characteristic of the emerging capitalist world system since the sixteenth century. However, today globalization has reached a qualitatively new stage because of its linkage with the other processes outlined earlier.

For Held, the consequence for nation–states, including new ones thrown up by globalization, is that their operation in an

> ever more complex international system both limits their autonomy and infringes ever more upon their sovereignty. Any conception of sovereignty which interprets it as an illimitable form of public power is undermined. Sovereignty itself has to be conceived today as already divided among a number of agencies national international and transnational and limited by the very nature of this plurality.[22]

However, as I pointed out earlier, it is important to note that this process of the loss of sovereignty is very much a double–edged one. A shift from a bipolar to a multipolar world can lead to greater autonomy for some states as it has increased the scope for action on the part of regional powers (while also intensifying attempts by core powers to regulate how this process occurs through the U.N. and other fora). The greater role of Nigeria, for example, in Western African regional security structures has been made possible by the end of the Cold War. Yet, say with the prod of nuclear proliferation, globalization could be associated with quite troubling independent actions by states in a multipolar power system.

A second consequence of these new times for the nation–state is that processes related to globalization are not politically neutral but are linked to the spread of liberal capitalism and to its likely global hegemony (the basis of Fukuyama's "end of history" thesis). In so far as this occurs, as liberal theorists like Fukuyama have been right to stress, war and military power will decline as mechanisms for resolving disputes amongst liberal capitalist societies. They will increasingly be confined to the interface between the capitalist and noncapitalist world. Moreover, this interface will no longer be structured by a bipolar superpower conflict; it will develop in a context where the communist road to modernity has lost legitimacy. This opens up the prospect not only for the advances of liberal democracy but also for conflicts amongst regional powers, as predator nations try to take advantage of the interregnum between a bipolar world and a fully peaceful globalized world. This prospect in turn increases the possibility of forming some kind of collective security arrangements, probably through the U.N. and underpinned by the military might of the one remaining superpower. While war and military power will be confined to the interface between the capitalist and noncapitalist world, it is certainly questionable that it will be easy for the developing world to tread the road to liberal democracy. As Max Weber and Alexis de Tocqueville warned us long ago, democracy has both a liberal *and* an authoritarian face and it is perhaps the latter one that is most often shown, as we may well discover in Eastern Europe. In any case, concerning the relations between liberal capitalist and nonliberal capitalist worlds, particularly those parts of the world which control scarce resources crucial for the survival of industrial societies, the end of the Cold War poses a serious question, namely, whether it has merely opened a new, far from trouble free, era of conflicts between North and South and amongst the states of the South, as advanced industrial societies seek to preserve their security through military means.

The Consequences of New Times for the Military Establishments of Advanced Industrial States

These developments provide the context within which we can finally consider the consequences of these "new times" for the organization of military power in the modern Western nation–state. Here we have to consider in particular the shift from a force in being to a cadre–reserve system which Moskos defines as a central part of the development of a warless society. Remember that Moskos is not predicting a world without war. Moskos is fairly sanguine about the persistence of regional conflicts, wars within countries, violent ethnic strife often connected with great population movements, and the likelihood that the major powers will be drawn into these conflicts. Yet he is suggesting that war "at least between super–powers and major European powers is no longer the principal much less inevitable mode of conflict resolution."[23] In this context, the recent Gulf crisis is merely a short-term hiccough interrupting a deep-seated process of "downsizing" and restructuring of armed forces observable throughout the Western world.

Specifically, Moskos expects the armed forces of advanced industrial states to be radically transformed by these larger social and political changes. With no need to be on continuous alert to fight global war, active duty forces will shrink dramatically to a well paid professional cadre which will train less well compensated long–term reservists. A militia and territorial defense system will evolve. The management–technician will be supplemented by the soldier–scholar as the officers corps becomes part of the general advisory service to the state on matters of international security. Indeed the warless society will give the military profession more time to study the requirements of their role which will have become much more broadly defined to encompass war fighting, peacekeeping, and aid to the civil authorities. Thus education and expertise in security studies, broadly defined, will become more important than a narrow focus on war fighting, strategy, and tactics. At the same time, in so far as states persist with the practice of conscription, civilian service will become a common alternative to service in the armed forces.

These developments will be associated with a blurring of the division between the different components of the national security system. As Martin Edmonds has argued, internal regional and ethnic strife has already undermined the traditional military/police divide in many states and this situation is likely to persist in both core and peripheral nation–states.[24] This "blurring" process, as I call it, will continue as armed forces take on not only more policing functions but also nonmilitary tasks such as civilian disaster management. Moskos also contends that the public will become more skeptical about the utility of armed forces. Furthermore, the services will increasingly have their work cut out to preserve their share of public expenditure against the competing claims of the ecological or environmental lobby, health, transport and other areas with greater legitimacy in public perceptions.

Thus, the post–Cold War armed forces face a perplexing scenario: one comprising "threat complexity"—that is, a very wide range of security risks which are difficult to prioritize—and a corresponding mixture of missions, ranging from high intensity war fighting to low intensity conflicts and peacekeeping operations. There is the added complication of being required to provide instant reaction forces to deal with security risks that are serious but fall short of major threats to national security and at the same time to provide the means of maintaining longer-term capacities to fight a major interstate war. Furthermore, these more complex tasks have to be performed within the context of a shrinking defense budget and a public more skeptical about the military establishment, seeking value for money from the defense sector, and likely to press for a further shift of resources to the nonmilitary state budget.

I would argue that this public skepticism (or at least indifference) and shift in priorities is actually part of a broader process of the decline of professionalism both military and nonmilitary.

I have argued elsewhere that the process of globalization can also be understood in terms of a shift in the focus of capitalism from within the nation–state to the global economy.[25] For example, the de–industrialization of certain sectors of advanced industrial societies can only be understood in terms of the industrialization of sectors of other societies which are more competitive from the standpoint of international capitalist enterprises; and their firms can and often do use the technologies of late modernity to realize their economic ambitions. Those in control of the states of modern capitalism thus have to provide the means of ensuring that a reasonable portion of globalized economic production and investment is persuaded to locate in their territory rather than elsewhere.

The age of neoliberal capitalism—the dominant form of globalized capitalism in these new times (especially in the Anglo–American context) through its emphasis on markets and its questioning of the legitimacy of collective state provision of goods and services—undermines the traditional power base of all forms of professionalism. Indeed it questions the vested interests of all "producer groups." The decline of professionalism can be discussed in terms of four dimensions:[26]

First, *a decline in monopoly power* to provide a service to clients; for instance, the opening up of a sector to competition from other producer groups, as with the opening up of property conveyance to groups other than solicitors.

Second, *an erosion of the distinctive culture or way of life* of a profession; for example, the increasing difficulty experienced by professions in resisting the imposition of the idea that efficiency in their world can be computed with performance indicators and the like in much the same way as in other areas of the division of labor. More broadly, this can be recognized as the "just another job" syndrome which has affected military service.[27]

Third, *an assertion of client power* over the professional group, particularly through the state and large business concerns; for example, see the struggles between the medical profession and the financial managers of state and private health enterprises. This client power is often asserted with the aid of other professions, most particularly accountants and related financial analysts.

And, finally, *a relative social devaluation of a profession* in the eyes of the public, discussed immediately below.

For the professions generally, the high point of their power and prestige was during the age of organized capitalism, stretching from the end of the nineteenth century to the 1970s, and it was during this period too that the public sector was both a seemingly ever–increasing consumer of national resources and regarded as having high legitimacy. With their emphasis on the values of altruism and service, the professions found a supporting cultural and structural system of support. However, this social base of the professions has been undermined by the development of neoliberal capitalism. For example, as Nikolas Rose has argued recently, in relation to the welfare state, these public bureaucracies have been subjected to attack by neoconservatism, neoliberalism and, from the left, various civil rights movements.[28] All of these have, in different ways, sought to defend the rights of the "autonomous family." Under neoliberal arrangements, there is an ongoing shift in the focus of the organization of welfare services away from the ministrations of subject citizens by monopolistic public welfare bureaucracies wielding producer power, towards a system of more self–initiated contracts between individual consumers—the "responsible family"—and state licensed but more competitive, rather than monopolistic professional and expert, producers. This creation of a market–led enterprise culture within a diminished and considerably delegitimated public sector is affecting not just the welfare state, but education as well—witness the establishment of devolved budgeting and bidding in British universities.

In the case of the modern military profession, we already have literature which is directly relevant to the theme of professional decline. This literature can be used to link the first and third of the criteria identified above, namely the decline of monopoly and the assertion of client power. For example, Cathy Downes has charted the ways in which the role of civilian experts in defense management has undermined the monopoly power of the military in delivering military advice/services to its client the state.[29] This assertion of civilian power should be located not only in the geopolitical and military context of the development of nuclear weapons and the related civilian political distrust of relying exclusively on the military for advice as to their development, deployment and use; it is related also to financial considerations and the domestic politics of defense budgets. For, as Martin Edmonds has argued, in many contemporary states, as a consequence of the emergent mismatch between the commitments of the national security system and the financial resources that governments feel able and willing to provide for it, more effort has been put into devising financial and administrative means of obtaining greater value for money from the system.[30] The emergence of neoliberal capitalism analyzed above, has given added impetus to this trend. Many modern states, including Britain

> manifest a desire for improved efficiency and expanded central financial control over defence spending. The pressure of spiraling defense costs for modern weaponry, full–time professional forces, and the requirements to maintain minimum levels of preparedness have necessitated certain structural changes within central organizations of defense. Furthermore, the emphasis on administrative and planning efficiency has resulted, to a varying extent in different countries, in the introduction of combined and joint planning staffs, and a heavy emphasis on the use and application of managerial and analytic techniques.[31]

These changes have involved a centralization of administrative power in this sector of the national security system together with a further subordination of senior military officers to civilians—either politicians or bureaucratic financial controllers. Indeed, what we are witnessing in the area of defense is the impact of neoliberal capitalism and related political strategies on the organized "producer power" of the military profession.

This is the broader political and financial context in which one should understand the emergence of smaller more flexible armed forces of the core countries; in other words, these emergent force structures are not simply the products of dramatic shifts in the geostrategic environment, crucial though these are. (Thus I would revise Moskos's warless society thesis which treats change in military organization as hinging on the risk of war as an explanatory factor and underplays the internal societal factors involved in the transformation of armed forces; an ironic observation of the present writer given my roots in sociological realism!). These financial constraints will also lead to development of systems of "force re–generation." Means must be found for reconstituting larger more capable forces should the need arise, but meanwhile smaller forces, able to deal with a wide range of less serious but immediately more likely threats, are deployed. Similarly, the advanced industrial societies will be predisposed to maintaining research and development in the most advanced weapons systems and technologies, but will not become committed to extensive production and deployment. In addition, smaller, more flexible forces will, in consequence of their multiple tasks, be likely to seek a more educated workforce: the armed forces will therefore become rather like another "flexible firm." One implication of these arguments is that, contrary to Moskos's hypothesis, the new armed forces and their officer corps, for financial reasons, will be extremely busy because they will be expected to do more and more varied tasks with fewer resources.

There is another issue: the military effectiveness of force structures organized along the lines of the cadre–reserve system. While such forces might be appropriate for demilitarized liberal societies, how will they fend (alone or as part of multilateral or even multinational forces) against nonliberal regimes with substantial military assets. Critical here are the capabilities of the reserves.

It is clear that it will be impossible, for financial reasons, and unnecessary given the warning times now preceding serious threats, for armed forces to have all specializations required for modern war in the regular order of battle. Indeed, ever since the mid–1970s, even though the U.S. had shifted from a mass armed force to a large force in being, for financial as well as political reasons, substantial elements of military capacity (indeed, half of the total military strength of the army) were placed in the reserve formations. This entailed something of a reversion to the old mobilization model, albeit on a volunteer basis.[32] While in the future, forces will be much smaller, the diminution of the threat and budgetary constraints will require continued and even greater reliance on reserve forces.

While reliable, high-quality reserve forces will be of importance in the future, there are nontrivial problems to be addressed about how such forces are to be raised and organized. In Britain, a recent defense White Paper reiterates the "great importance" that HMG attaches to the volunteer and regular reserves of all three services, "who will continue to have a key role to play." Plans for future restructuring envisage the development of "a manpower structure from 1995 onwards [that is] best matched to anticipated tasks, reduced readiness requirements and extended warning and preparation time."[33] As we shall see presently, current changes being planned for the reserve organization, as a result of a Ministry of Defense study, will involve something of a shift towards a cadre–reserve model. Of course, the reserves have played a crucial part in the British armed services for many years. The ability effectively to mobilize reserves to reinforce the British Army of the Rhine over the last forty years, which has been successfully demonstrated in several major exercises, has been crucial to the defense of the central front, and thus has contributed a key component of flexible response and forward deployment. Reservists also played an important role in the recent Gulf War, particularly in the medical and other specialist technical areas. In the post–Cold War context, despite some important changes, reservists will continue to play an important role, although how important is a controversial matter.

This is a complex issue involving the effects of changes in warning time on the forces, the political will to call out the reserves in time of crisis, and the technical capacities of the reserves to meet the requirements of likely conflicts in the future. Reserves comprise volunteers on the one hand and ex–regulars with a call–out commitment on the other. Clearly, although there is a transition period, the ex–regular pool dwindles as the size of the regular forces decreases. Any future expansion of the reserves, therefore, will have to rely largely on the volunteer sector. Yet, if the services are to attract young people to the volunteer reserves, more will need to be done to provide better quality conditions of service and training facilities. One possibility would be to introduce compulsory service, with or without the option of civilian community service, although this raises complex political issues which have not yet been given a wide airing, at least not in Britain. In any case, whatever policy is adopted in relation to the future use of the reserves, there will need to be revised legislation facilitating their early call out, otherwise it will not be possible to put them to their best possible use.[34]

Assuming the forces will be able to recruit sufficient numbers of volunteer reservists to supplement a smaller regular force, the question arises, what will they be used for in the post–Cold War world? Traditionally reservists have been used in the highly skilled areas where expertise is built up over years such as in the mine countermeasures force in the Royal Navy, or in simpler but nevertheless fully operational areas such as the infantry. With short warning times, limited peacetime training and ever more sophisticated equipment, it has proved impossible in the past to use volunteer reserves to a significant degree in manning major equipment in all three services. This point even applies to the ex–regular who fairly quickly gets out of date and loses touch with the relevant skills.

In the light of the new security climate and shifts in defense policy, it is possible that this situation could change. Assuming a continued process of demilitarization in Europe, certain equipment could be stored. One related possibility is that these could be reactivated in a longer warning time whilst, in parallel, the relevant reserves could be called up to man them. On this issue it is likely that different equipment would call for different answers in terms of warning times and reserve call outs. Even assuming that reservists called out in this way would be sufficiently able technically to perform the roles required of them, these arrangements would require appropriate investment in storage maintenance and training capacity. Both maintenance and training would need to "tick over" in peacetime and have a credible surge capability in a time of tension. Reservists should not simply be viewed as a cheap supplement to a smaller regular force if the principles of credibility and military effectiveness are to be sustained.

The Gulf conflict has revealed the importance of having a professional and well–prepared reaction force to meet emergencies of this kind in the future. For military and political reasons, it is difficult to imagine reserve forces, at least as traditionally organized and trained, playing a key role in these future arrangements. Yet, for serious large–scale emergencies, it would be imprudent not to provide a reserve and infrastructure surge capacity which can be triggered when necessary.

This points to a likely development not of a cadre–reserve system as such but rather to a dual military structure. First, there must be a professional reaction force, supported by specialists and a core of properly ready reservists. For economic reasons, smaller regular forces in the post–Cold War context will have to come to terms with the fact that it is impossible to have all the military specializations needed by a modern armed force located in the regular formations. At the same time, regular forces, especially rapid reaction forces, will have to be persuaded that working with reservists need not entail working with "amateurs." Reservists earmarked for early use with rapid reaction forces need to be properly trained and equipped, and shown to be capable of performing effectively on an increasingly complex battlefield. For this reason, such reservists are most likely to be found in those occupational sectors where there is a close convergence between military skill demands and the skills performed in civilian society—among engineers, transport and logistics specialists, medical doctors and health service personnel, and so on. (It might still be the case, of course, that the political sensitivity associated with deploying reservists will make governments—as in the U.K.—reluctant to use them, even if their military effectiveness can be proved. Yet the financial pressures to "cadreize" the armed forces, that is to rely more heavily on reserves, will be immense. Such are the conflicting pressures of politics and economics on decisions about the nature and use of armed forces in these new times!) This professional reaction force, with appropriate reserve attachments, would be a component part of a multinational defense system for use in Europe or elsewhere to cope with security problems at the interface between modern and modernizing societies. These crises are likely but not serious in terms of constituting a threat to the homeland. Second, a more residual reserve force is required. It would be linked to an infrastructure of surge capacity to meet problems which do constitute serious threats (however unlikely) to the homeland, and manned by reservists who could be prepared during the period provided by longer warning times expected before any such large–scale conflict. As Antony Beevor has graphically argued, this would actually look like a three–leveled cake with professional regular battalions supplemented by a ready reserve and a residual reserve.[35] The model could be linked to the more regional basing of the British army and to a greater role for specialization. Assuming sufficient funds and political will, this would be a force structure that sought to deal with three objectives in these new times of ours: first, to provide a credible response to threat complexity in terms of the mix of war fighting and peace keeping; second, to be able to respond to short and long term security risks appropriately; and third, to mesh financial constraints with strategic prudence.

In building such a dual military structure, four key issues relating to the reserves will have to be borne in mind. One is recruiting and retaining personnel in the context of a public perceptions of a declining threat. Another is that the experience of the regular reserves may well prove to be an important resource which up to now has been under–utilized due to shortages of funds. Third, using reserves is a major political gesture and this may well not be the best way to respond to crises which do not threaten vital national security interests. Finally, if shifts in the regular/reserve balance are allowed to be driven largely by financial pressures and insufficient resources are devoted to the revised arrangements, then a cheaper war machine may be purchased at the cost of military effectiveness, a situation that it will not be easy to put right in the next major military emergency.

Conclusions: War and Peace in the Core and Periphery

The overall trend towards what has been termed a cadre–reserve system as part of a warless society seems to me to be a fruitful hypothesis when looking at the current prospects for the reorganization of Western armed forces or indeed of the armed forces of any capitalist liberal democracy. However, we must be careful not to become entrapped in evolutionary or "unfolding" models of contemporary history. Moskos accepts that his thesis on the social organization of military power hinges on a decline in the prospect of war as a means of resolving conflicts amongst capitalist liberal democracies and on the implicit claim that interventions by these powers in so–called third world disputes can be disentangled from serious confrontations amongst major and/or superpowers. Furthermore, it seems to me that his case also rests on the assumption that, when armed forces do become involved in third world conflicts, the conflicts can be contained as "small wars," rather like in the colonial past and the more recent brush–fire war era.

Let us assume for now a continuing trend towards the global growth of liberal capitalism, part and parcel of which would be a successful prosecution of liberal "reform" in what replaces the Soviet Union, the emergence of Eastern European security structures which avoid the precedent being established in (what was) Yugoslavia, and a shift from dictatorship to democracy in the Middle East in the post–Gulf War context. These developments could be construed as gradual steps towards the realization of a system of collective security centered on the U.N. If so, one can envisage further changes in the social organization of military power, in particular the development of multinational forces, which would represent a broader extension of the sort of arrangements discussed in the context of the possible emergence of a pan–European security system.

But what if the spread of capitalist liberal democracy does not take place as envisaged in this developmental image? What if, from the standpoint of the developed world, we are facing a scenario of wars with or amongst major regional "have-not" powers? What if the current Gulf War is not a "one off" so to speak? In regard to these threats, the difficulties of developing multinational (not just multilateral) forces as an arm of the U.N. will be lessened somewhat by the end of the Cold War and the prospect of a system of collective security underwritten by the United States. This would be a Pax Americana to mediate security between the developed and developing world. The

United States, as the only superpower left, could play this role, but only in so far as the relatively nonmilitarized major economic powers such as Germany (and the rest of the European Community) and Japan underwrite it economically.

Nevertheless, threats to capitalist democracy from regional dictatorships may well persist. After all, with the industrialization of war, it is not that difficult for middling economic powers to acquire the military means of becoming a serious regional military threat. And it is naive to suppose that these threats will not create serious divisions of interest and conflict amongst the major capitalist powers themselves. As Will Hutton suggested recently in the context of conflicts over GATT and agriculture, let alone over the Gulf crisis, we could easily witness a fairly rapid shift from a bipolar through a short–lived unipolar world—with a weak U.S. at the top—to a "multi–polar world of competing trade, currency and security blocs."[36] While those conflicts might be insufficient to provide the basis of a *casus belli* between the capitalist powers, it is not unreasonable to suppose that an unwillingness to trust the guarantees of the United States, or of course a reluctance on the part of the United States to give them, might provoke one or more blocs to establish their own revised military security systems adequate to meet the prospect of changing regional security threats. At the same time, it also seems reasonable to suppose that in this global age such arrangements would not be developed and applied in isolation from the U.N.

One can thus identify two alternative global security structures in which multinational cadre–reserve type systems might be located: the U.N. global police model and the regional bloc security model. We could say that the social organization of military power transcended the nation–state to the extent that either of these emergent alternative security systems was based, not just on multilateral but, on multinational forces structured along the lines of a cadre–reserve model. Given the persistence of Machiavellian factors in the pursuit of power, we should recognize however that both scenarios fall far short of delivering the global security and enduring peace foreseen by the utopian, progressive models of social change which we have inherited from nineteenth–century social theory.

Notes

1. Christopher Dandeker, "Armed Forces and Society Research in Great Britain: The Prospects of Military Sociology," *Forum*, 8 (1989): 1–34.
2. Martin Shaw, *Post–Military Society* (London: Polity Press, 1992).

3. John Mearsheimer, "Back to the Future: Instability in Europe after the Cold War," *International Security* 15 (1990): 5–56.
4. See Anthony Giddens, *Nation–State and Violence* (London: Polity Press, 1985); Michael Mann, *War States and Capitalism* (London: Basil, Blackwell, 1989); John A. Hall, *Powers and Liberties: The Causes and Consequences of the Rise of the West* (London: Pelican, 1986); Martin Shaw, *War, State and Society* (London: Macmillan, 1985); and Christopher Dandeker, *Surveillance, Power, and Modernity* (London: Polity Press, 1990).
5. Francis Fukuyama, "The End of History?" *National Interest* (Summer 1989): 3–18.
6. Charles C. Moskos, "Armed Forces in a Warless Society," paper prepared for the British Military Studies Group, King's College, London, 1991.
7. Ibid.
8. Morris Janowitz, *The Professional Soldier* (New York: The Free Press, 1971) and Jacques Van Doorn, "The Decline of the Mass Army in the West: General Reflections," *Armed Forces and Society* 1 (1975): 147–157.
9. James Burk, "The Decline of Mass Armed Forces and Compulsory Military Service," *Defense Analysis* 8 (1992): 45–59.
10. Ibid.; see also Morris Janowitz, *The Professional Soldier*, 417–426.
11. James Burk, "The Decline of Mass Armed Forces"; see also Philipe Manigart, "The Decline of the Mass Armed Force in Belgium," *Forum* 9 (1990): 37–64.
12. See Jacques Van Doorn, "The Decline of the Mass Army in the West: General Reflections," *Armed Forces and Society* 1 (1975): 147–157; Gywn Harries–Jenkins, *From Conscription to Volunteer Armies*, Adelphi Paper No. 103 (London: International Institute for Strategic Studies, 1973); Morris Janowitz, *The Professional Soldier*, 417–426; and Bernard Boëne, "How Unique Should the Military Be? A Review of Representative Literature and Outline of a Synthetic Formulation," *European Journal of Sociology* 31 (1990): 3–59, esp. 14–27.
13. Philipe Manigart, "The Decline of the Mass Armed Force in Belgium"; James Burk, "National Attachments and the Decline of the Mass Armed Force," *Journal of Political and Military Sociology* 17 (Spring 1989): 65–81.
14. Anthony Giddens, *The Consequences of Modernity* (London: Polity Press, 1991), 55–78.
15. My discussion of this topic is indebted not only to the important work of Anthony Giddens, but also to as yet unpublished ideas of Wilfred Von Bredow on the question of globalization and modernity, some of which I have adapted for the analysis presented here.
16. Anthony Giddens, *The Consequences of Modernity*, 64.
17. A New World Order: To the Victors, the Spoils—and the Headaches," *Economist* (September 28, 1991): 25–27.
18. Speech of H.M. Secretary of State for Defense M. Rifkind to the Royal United Services Institution, London, February 20, 1993.
19. Martin Van Creveld, *On Future War* (London: Brassey's, 1991).
20. See P. Alter, *Nationalism* (London: Edward Arnold, 1989), 92–152.
21. David Held, "Farewell the Nation State?" *Marxism Today* (December 1988): 12–17.
22. Ibid., 16.
23. Charles Moskos, "Armed Forces in a Warless Society."
24. Martin Edmonds, *Armed Services and Society* (Leicester: Leicester University Press, 1988), 113–160.
25. Christopher Dandeker and Paul N. P. Watts, "The Rise and Decline of the Military Profession," *Forum* 12: 75–105.

26. Ibid.
27. Charles C. Moskos and Frank R Wood, eds., *The Military: More than Just a Job?* (Washington, D.C.: Brassey's, 1988).
28. Nikolas Rose, *Governing the Soul* (London: Routledge, 1990).
29. Cathy J. Downes, "To Be or Not To Be a Profession: The Military Case," *Defense Analysis* 1 (1985): 147–171.
30. Martin Edmonds, ed., *Central Organizations of Defense* (Boulder, Colo.: Westview Press, 1985).
31. Ibid., 14.
32. David R. Segal, *Recruiting for Uncle Sam: Citizenship and Military Manpower Policy* (Lawrence: University Press of Kansas, 1989), 152–156.
33. Ministry of Defense, *Defense Estimates* (London: HMG Printing Office, 1991), 42.
34. Christopher Dandeker and James Higgs, "Armed Forces after the Cold War: The Personnel Implications," paper prepared for the British Military Studies Group, King's College, London, 1991.
35. Antony Beevor, *Inside the British Army*, revised ed. (London: Corgi, 1991), 350.
36. *Guardian*, January 23, 1991, 15.

6

The Postmodern Military

Charles C. Moskos and James Burk

Students of military history have never embraced the stereotypical view that modern military organization is a rigid, hierarchical, and unchanging bureaucracy. Whether national armies and navies have adapted well or poorly to the changing environments of war and peace, they have rarely remained unchanged organizationally. The history of modern military organization is a history of flux. The critical problem for historians and social scientists and for policymakers is to discern the underlying patterns of change and their significance for defining the military's social role and evaluating its capacity for fighting wars. The task, unfortunately, is far from easy.

While general factors are at work, it is impossible to explain the changing format of modern armed forces in terms of single factors. We may examine, for example, the organizational effects of modern technological innovations, from tanks and machine guns to thermonuclear weapons and precision guided missiles. Or we may, more ambitiously, notice how the shift from agrarian to industrial economic organization vastly complicated wartime logistical operations, freeing campaigns from many natural constraints, but adding new social constraints, and blurring the distinction between civilian and military functions. These and other factors are obviously important determinants affecting what forms military organizations might assume. But none is all–determinative. Adequate discussion of military change must take a wide variety of factors into account, including technological innovation and economic organization to be sure, but encompassing other political, social, and cultural factors as well.

The aim is to distinguish lasting from ephemeral change in military organization. For this purpose, we undertake a systemic institutional analysis, a perspective that tries to account for the organizational importance of long–term historical developments. Harold Lasswell's classic essay on the "garrison state" remains the model to be followed.[1] Adopting such a perspective, we can identify critical periods of transition in military organization, when ideas about how militaries should be organized and what they should do were fundamentally transformed. It is commonly argued, for instance, that the end of the eighteenth century, marked by the American and French Revolutions, was such a period, a time when the idea of the modern mass army, based on the conscription of citizens fighting for the nation–state, was conceived and, to a limited—at first, actually, to a quite limited—degree, displaced the professional armies, fighting for absolute kings, which dominated Europe for 150 years following the Peace of Westphalia.[2] While not born in full bloom, the idea of the mass armed force gradually gained adherents, notably after 1870, and was fully developed during the two world wars of the twentieth century.

The question we would address in this chapter is whether now is another similar period of transition and, if so, what is the new idea of military organization and purpose. The end of the Cold War has raised the question with special urgency. But in fact the question has been guiding sociological studies of military organization for over two decades. Our working hypothesis is that we are indeed in a period of transition away from the "modern" mass army, characteristic of the age of nationalism, to a "postmodern" military, adapted to a newly forming world–system in which nationalism is constrained by the rise of global social organizations. Much of our analysis will consist of a comparison of these two types of military organization along a variety of dimensions. Our goal is to clarify as much as possible what we mean by the idea of a postmodern military organization. We shall also, briefly, consider the likely prospects of the movement towards a postmodern format for civil–military relations. Yet to begin, we should ask why it is reasonable to regard the present as a period of transition for military organization.

A Period of Transition

The end of the Cold War concluded this century's era of world war which was characterized structurally by deep–seated conflict and overt hostility between two shifting blocs of empires and nation–states. We may debate, as William Thompson has done, whether the end to this round of global war is permanent or only a temporary respite.[3] Yet, whatever we decide, there are many reasons to believe that we are entering a period of transition (and may perhaps have been entering it over the last several decades) in which the modern forms of military organization and war are giving way to new, postmodern forms. Moreover the change taking place is not simply confined to the military or to the realm of war. On the contrary, we are dealing with a sweeping reorganization of modern societies.

There are several indicators that sweeping change is taking place. Of special relevance for us is the relative weakening of central forms of social organization which have been hallmarks of the modern age: the nation–state and national markets, democratic citizenship, and the mass armed force. Or, put in positive terms, the substantial growth of global social organizations has altered the conditions under which modern nation–states can expect to exercise their power, maintain the loyalty of their citizens, or raise and deploy their military might. As important, these structural changes are accompanied by a "cultural shift" in public attitudes and opinion. Old verities are questioned, rather than accepted. There are few if any overarching authorities to which people are willing to defer. There is a shrinking consensus about what values constitute the public good, nor even much confidence that we know how, by the use of reason, to determine what the public good might be. The eighteenth century's faith in reason, the nineteenth century's faith in the nation–state, and our own century's confidence in science and technology have all lost their hold on our imagination, despite their considerable accomplishments. Now is a time of general uneasiness.[4] It is precisely this transition which social theorists mean when they refer to our's as a postmodern age.

The postmodern movement in social theory began, in the late 1960s and early 1970s, as a form of literary criticism. The concern was with the meaning of words and with their relation to reality. A central claim of the movement, to state the matter in oversimplified terms, is that words, rather than what we say words refer to, define what we think of as reality.[5] The idea that the words, or more generally that the symbols, we use to comprehend the world can in fact provide us with an accurate (the true) picture of the world is, accordingly, an illusion, the illusion of modern realism. But, as Jean–Francois Lyotard understands it, this is not by any means an innocent illusion.[6] Its aim is to provide us with false assurance that we can have certain knowledge of a world that is fundamentally uncertain, to give us a false sense of power, security, and control. Yet what it leads to in fact, as any quest for certainty is bound to do, is a reign of terror. So, to be concrete, the quest for ethnic purity pursued to fulfill the nationalist dream of a homogenous population, has led to the Holocaust and to "ethnic cleansing."[7] For Lyotard, postmodernism is a stance which we deliberately take against this modern illusion. It seeks to make clear that there is no whole, no pure totality, no reconciliation of the concept and the sensible in the world. And if not, he hopes, there can be no resumption of the totalitarian horrors for which the modern era has become too well known.

As Lyotard's argument suggests, postmodern theorists are skeptical about the existence of any ultimate standards of knowledge or morals that we might use to support judgments about what is happening in the world. Their skepticism leads them to criticize hierarchy, grand narratives about national traditions, unitary notions of authority, and the bureaucratic imposition of official values.[8] But it also leads to nihilism, where we (for reasons beyond the scope of this chapter) prefer not to go.[9] We would be mistaken, however, to suppose that postmodernism refers to nothing more than an idle intellectualist critique of modern society. What supports postmodern claims and makes the theory noteworthy, despite all its pretensions, are real transformations shaking contemporary social organization. Nation–states and their institutions are becoming more fragmented and decentralized, as they try to respond effectively to an increasingly global social order and yet also to retain their local power and relevance. The results of this may be shown, as James Kurth has argued recently, by examining the strains and changes in cultural, economic, and military organization.[10]

Referring specifically to the United States, Kurth claims, the modern era is at an end, that the United States is no longer a nation–state, as we have traditionally understood the term. Historically, the nation–state entailed three elements: (1) a common culture, learned ideal–typically through a system of public schools and so entailing the spread of literacy; (2) a unitary bureaucratic state, governing over the territory and, again ideal–typically, defending the territory with a large conventional army, based on mass conscription; and (3) a common market within the state to which economic endeavors were primarily oriented. When Kurth examines recent social trends along all three of these dimensions, he concludes that the United States is better described as a postmodern society, or a multicultural regime.

American cultural organization is no longer dominated by the school system offering instruction in a national culture. It is dominated rather by the (often electronic) media of mass entertainment which are oriented to a transnational but highly fragmented pop culture. In the same way American economic organization is no longer characterized by large industries engaged in mass production and distribution for a national market. On the contrary, economic organization has grown increasingly specialized to provide financial and managerial direction to large–scale multinational enterprises or, at the lower end of the scale, to provide a plethora of retail services that can only be produced and consumed at the local level.[11] And finally, militarily, the ideal–typical form of a national military, associated with universal male conscription, masculine virtues, and national patriotism, has been transformed into a "high–tech" professional armed force, providing military power for temporary international coalitions. In this context, the meaning of explicitly national defense has grown very elastic. Before turning to explore in greater detail the character of this change in military organization, it is worthwhile illustrating the effects of these changes on the institution of citizenship.

Citizenship is a distinctively modern institution. As Philip Wexler has argued, the practice of citizenship depends on the modern liberal conception of an autonomous individual, capable of free choice and self–regulation, participating in public affairs as he or she decides, and participating with every expectation of exerting influence on the outcome of political decisions.[12] Yet none of this is possible Wexler claims in a postmodern society, and not for the reasons that Joseph Schumpeter made against the classical theory of democracy, which focused on the real politics of electoral struggle among contending political parties.[13] Rather the practice of citizenship has been turned into a mere "simulation," a sign with no reference. Wexler quotes, with approval, the observation of Timothy Luke that "elections, in fact, are now commodified and packaged modes of democracy" and that "citizenship is now like being a fan, who votes favorably for media products by purchasing them, extolling their virtues, or wearing their iconic packaging on one's bill cap or tee shirt."[14] Anything more is too much to expect. The boundaries creating separate sovereign domains (which made the modern nation–state possible at the global level, and made the autonomous individual possible within the state) are blurred and disintegrated by the transnational character of cultural, economic, and even military organization. And so the sense of identity with and loyalty to the nation–state is "decomposed" in postmodern society.

The central point, for our analysis, is that postmodern society is distinguished from modern society by the transition from certainty to radical uncertainty about the meaning or purpose of central roles and institutions. In this situation, we cannot easily judge the relative importance of various collective activities. We doubt whether there are or can be any common standards or rules to apply that could establish the content and scope of social rights and obligations. Let's turn now to consider the consequences of this transition for military organization and civil–military relations.

From Modern to Postmodern Military Organization

Changes in military organization reflect, as they sometimes effect, large–scale changes in social organization. Comprehensive analysis of these reciprocal relations requires a clear specification of the dimensions along which change is expected to occur. As a practical matter, we rely on typologies of military organization to accomplish the task, even though we are fully aware that any typology does an injustice to reality. Drawing heavily on the historical experience of Western European nations, to include the United States, it is possible to describe and contrast modern and postmodern military organizations, and to speculate about the factors facilitating movement from one type to the other.

For this purpose we posit three types of relations between the military and society. The first is the *modern* type which dates, roughly speaking, from the late eighteenth century to the mid–twentieth century. The second is the *postmodern* type which we believe is emerging in the present and will persist into the indefinite future. A third, the *late modern* type, is added to help explain the transition from modern to postmodern military organization. It dates from the mid–twentieth century to the early 1990s. Although the typology draws heavily from Western experience, the essential differences between armed forces and society in these three types are couched in general terms suitable for broad cross–national analysis.

The basic argument about each type is summarized in table 6.1. Our concern is to grasp the whole, to place the salient facts within a framework that will enable us to study the main trends of institutional development in military organization. The typology, in other words, is offered as a guide to systematize current research findings. We must avoid using it mechanistically to bring artificial closure to our thinking about these matters. Its use, rather, is to help bring focus to ongoing research and, if need be, to set the stage for revising the analytic framework we are about to present.

TABLE 6.1
Armed Forces and Postmodern Society

Armed Forces Variable	Early Modern (Pre–Cold War)	Late Modern (Cold War)	Postmodern (Post–Cold War)
Perceived Threat	Enemy invasion	Nuclear war	Subnational and nonmilitary
Force Structure	Mass army	Large professional army	Smaller professional army with reserves sharing missions
Public Attitude toward Military	Supportive	Ambivalent	Skeptical or apathetic
Impact on Defense Budget	Positive	Neutral	Negative
Organizational Tension	Service roles	Budget fights	New missions
Dominant Military Professional	Combat leader	Manager or technician	Soldier–statesmen; soldier–scholar
Civilian Employees	Minor component	Medium component	Major component
Women's role	Separate corps or excluded	Partial integration	Full integration
Spouse and Military Community	Integral part	Partial	Removed
Homosexuals in the Military	Punished	Discharged	Accepted
Conscientious Objection	Limited or prohibited	Permitted on routine basis	Subsumed in civilian Service

Changes in Threats and Threat Perception

We begin with the simple idea that the probability of war and the perception of threats shape the basic relation between armed forces and society.

One key difference between modern and postmodern societies lies in the character of the threats they face and the ways they perceive them. For modern states, the threat of enemy invasion of the homeland, or of close allies, is always a real possibility, a live option that has to be defended against. Indeed, Charles Tilly has recently argued that the rise of the modern nation–state in Europe rested on its success in meeting these threats.[15] Over the last two centuries, an important factor accounting for a state's survival, has been its ability to mobilize and deploy a mass armed force, relying typically on a system of compulsory military service to prepare its population for war. In the modern era, the threat of war, and so the justification for armed forces engaged in border defense, was always close at hand. Mobilization to meet this threat, at least since the end of the eighteenth century, was one of the main sources of nationalist fervor.

The situation for postmodern societies is quite different. The hallmark of a postmodern society is that the military threat to its survival becomes muted and blurred. No serious observer, of course, sees the imminent end of large–scale violence. The prevailing discussion of the termination of the Cold War, of a denuclearized Europe, and of conventional armed forces greatly reduced in size is not so naive as to mean the literal end of war. The future can expect to see regional conflicts, civil wars, violent ethnic strife, and major power military interventions to defend spheres of influence, just as the present and past have done. But the prospects for large–scale or global war, which many feared were high in the early 1980s, are not high now, nor is there any strong likelihood that they will rise in the foreseeable future. The events that rapidly developed between the United States and the Soviet Union and within Europe during the late 1980s led to a momentous, not an illusory, reduction in military tensions. Wars between developed countries in the West and formerly socialist countries are moving toward the realm of improbability.

We are, to be sure, struggling to assess the significance of this fact. Present trends toward democratization are greeted happily in the knowledge that parliamentary liberal democracies have not historically gone to war against one another. Yet uncritical optimism may not be warranted. It is not the movement from war to peace that defines the movement from modern to postmodern society. It is, rather, a movement from being relatively certain what war would mean and what victory could secure, of knowing who one's enemies were and whether they were defeated, to a situation in which all these matters are very far from clear. The transition from modern to postmodern society, understood in these terms, may be tied to the new threat of nuclear war in late modern societies, with all the complexities it introduced into calculations about the use of armed force for national defense. Martin Van

Creveld has raised the provocative thesis that war, while it may always be with us, has fundamentally changed its character.[16] We are moving away from the Clausewitizian trinity of the state, the army, and the people. Wars between nations, he thinks, will be replaced by intrastate warfare in which national boundaries will no longer hold a central place. Terrorism and ethnic violence will be the postmodern mode of war, involving tasks for which the conventional military is ill–prepared to cope. If this analysis is correct, we are witnessing the dawn of an era in which war between major powers is no longer the principal, much less inevitable, mode of conflict resolution. It is an era in which not just political and military elites, but the publics of advanced industrial nations display marked reluctance to become engaged in protracted, uncertain wars in parts of the world they regard as secondary to their concerns.

From this vantage point, even the Persian Gulf War may have to be seen as something other than a clear–cut demonstration of a modern military victory. In August 1991, a year after he invaded Kuwait and five months after he was defeated in battle, Saddam Hussein boasted about his famous victory over Baghdad radio.[17] The Iraqis, he claimed, were the true winners of the war because they remained defiant in the face of superior odds. It is true of course that Iraq was driven out of Kuwait. But *is* that the standard for judging who won or lost the war? The Gulf War was a quintessentially postmodern war insofar as no clear standard exists allowing everyone to agree on the answer to this question.

Changes in Force Structure

How will military organization, a central institution for promoting the development of the modern state, be affected by and, equally important, affect these developments? Answering the question requires that we avoid a one–sided emphasis on trends in strategic doctrine, weapons technology, and "correlation of forces," and that we also avoid one–sided emphasis on new trends in peace movements and the search for alternative means of conflict resolution. It requires rather that we focus on the institutional role and dynamics of the military in relation to the larger society.

The dominant change in military force structure is away from the mass armed force model, based on conscription, toward a smaller, voluntary, professional force that relies on reserve forces to accomplish its missions.[18] The change in format is closely related to changes in the missions armed forces have to perform, which is to say they are closely tied to the transformation of threats we have just discussed. Preparation to fight global war on the scale of the two world wars provided the central mission and justification for mass armed forces. In the late modern period, however, concern about conventional global war gave way to concern about the nuclear arms race, the consequences of fighting nuclear war, and the need to build institutions designed to prevent the outbreak of nuclear war. In a period during which war fighting on a large–scale is increasingly seen as a self–defeating strategy, the moral legitimacy of conscription wanes. Moreover, as expensive and complex weapons systems are added to military arsenals, the need for highly trained, long–term, and relatively well–paid professional forces emerges. These forces remained large in size throughout the Cold War era in order to sustain a credible deterrent defense. Yet as we move into the postmodern era, the meaning of deterrence is no longer the same, and what constitutes the military's mission is subject to revision.

It is instructive, here, to examine the various missions Western military forces have undertaken just since the end of the Gulf War in March, 1991. Western European and American forces have participated in over a score of deployments of various sizes and duration from the end of the Gulf War to the time of this writing (mid–1993). A list of these deployments is given in table 6.2. It includes Operation Provide Comfort for Kurdistan relief, Operation Sea Angel for flood relief in Bangladesh, a multinational peacekeeping force as well as relief missions in Yugoslavia and its successor states, rescue work following volcano eruptions in the Philippines and Italy, a multinational peacekeeping force in Cambodia and Somalia, rescuing of foreign nationals in Zaire and Chad, restoring domestic order in Los Angeles, and hurricane relief work in Florida. What these missions share in common is a new emphasis on nonwar fighting military missions. Indeed, the term "military humanitarianism" has been used to describe the emergent roles of the armed forces.

With these changes in threat perception and force structure, public attitudes toward the armed forces are subject to change as well. In the conventional modern nation–states of Europe, from the nineteenth through the first half of the twentieth century, public support of the military was linked with patriotism and nationalism and was quite high. The ability of Britain during the First World War to rely for so long on volunteers to raise its armed forces stands as an exemplary, if not altogether rational, policy that rested on this support. During the late modern period of cold war and nuclear standoff, public attitudes changed to view armed forces

TABLE 6.2
Post–Gulf Military Roles of Western Nations

Location	Date	Mission	Participants
Kurdistan "Operation Provide Comfort"	April–June, 1991	Refugee relief	12,000 U.S. forces, 11,000 coalition partners
Bangladesh "Operation Sea Angel"	May–June, 1991	Flood relief	8,000 U.S. Marines
Former Yugoslavia	July 1991–	Cease–fire monitors	E.C. civilian & military officials
Albanian "Operation Pelican"	July, 1991–	Relief work & port patrol	800 Italian military
Phillipines "Operation Fiery Vigil"	July, 1991	Mt. Inatubo volcano rescue	5,000 U.S. Navy & Marines
Western Sahara	September, 1991	Observer force	U.N. military with U.S. officers
Zaire	September, 1991	Rescue foreign nationals	French & Belgian troops with U.S. airlift
Cuba	November, 1991– May, 1992	Haitian refugee relief	U.S. military
Cambodia	December, 1991	Peacekeeping	Projected 16,000 U.N. force
Russia "Operation Provide Hope"	December, 1991– February, 1992	Food relief	Western & U.S. airlift
Chad	January, 1992	Rescue foreign nationals	French contingent
Former Yugoslavia "U.N. Protection Force	March, 1992–	Peacekeeping	1,500 force, projected 7,000; also W.E.U. and NATO naval deployment off shore
Italy "Operation Volcano Buster"	April, 1992	Mt. Etna volcano rescue	Small number of U.S. Marines and Navy forces
California "Joint Task Force Los Angeles"	May, 1992	Restore domestic order	8,000 U.S. Army & Marine Corps, 12,000 National Guard
Somalia	August, 1992	Famine relief	2,000 U.S. Marines off shore, small U.N. force with U.S. airlift
Florida "Joint Task Force Hurricane Andrew"	August– September, 1992	Diaster relief	21,000 U.S. Army, Air Force, Marines, & 6,000 National Guard
Iraq "Operation Southern Watch"	August, 1992–	Surveillance	U.S. Air Force, Navy
Hawaii	September, 1992	Hurricane Iniki disaster relief	National Guard with a small number of U.S. Marines & Air Force
Somalia "Operation Restore Hope"	December, 1992–	Peacekeeping	U.S. forces initially; U.N. force in spring, 1993

On–going Peacekeeping Forces

Year Formed	Force Name	Acronym
1964	U.N. Peacekeeping Force in Cyprus	UNFICYP
1974	U.N. Disengagement Observer Force: Israel/Syria	UNDOF
1978	U.N. Interim Force in Lebanon	UNIFIL
1982	Multinational Force and Observers: Sinai	MFO

with much more ambivalence than before. Still, the levels of mass support of the military during this period were higher than the levels of support offered by the more critical intellectual and cultural elites. This difference was especially evident in Western societies during the 1960s and 1970s. But in the postmodern era, following the Cold War, even the general public has become, if not outright skeptical, then certainly more apathetic about the military and military matters.

The impact of these shifts in public attitudes on military organization is perhaps most easily observed through trends in defense budget figures. Without any obvious need for armed forces to meet traditional threats against foreign invaders, military spending comes to be seen as hindering economic growth. Money spent on national defense is thought to reduce the availability of funds for state–sponsored benefits, notably in the areas of health care and education. While defense spending measured as a percent of gross national product was high in modern societies, in postmodern societies, military budgets shrink in response to political pressure for a "peace dividend." Not all force restructuring, in sum, can be understood in terms of military planning to meet changing threats in the world. Without firm public support, the armed forces of democratic countries are constrained by the amount and kinds of resources they are given to work with.[19]

As might be expected, the character of intraorganizational tensions changes as the military moves from the modern to the postmodern period. In the modern armed forces, intramilitary conflict primarily revolves around definitions of the roles of land, air, and sea forces and, later, includes contests over the tightening defense budget. In the post–Cold War era, along with continuing budget fights, whether the military should substitute other goals for its war fighting mission becomes a major topic of controversy. As our listing of post–Gulf War deployments indicates, attention shifts to such new roles as multinational reaction forces, peacekeeping missions, disaster relief, and the like. Other candidates for new missions include space and Arctic exploration, anti-drug trafficking, education and job training for deprived youth. At issue are questions, for instance, about which services should perform what roles, how combined operations in these new fields should be managed, and how these many varied missions should be ranked in order of importance, to decide how much attention and resources should be paid to each one.

These tensions are aggravated by the military professional's self–image as a specialist in violence, ready for combat. Yet even here we see patterns of change over time. Probably the most documented finding in military sociology tells how the dominant type of military professional shifts from the combat leader during the modern period of war readiness to the managerial technician in the late modern period of war deterrence. This shift was most clearly argued for in the American case by Morris Janowitz.[20] Yet the trend was international in scope, affecting the armed forces of all industrial nations, with local variations of course, as they moved through these two periods.[21] In the postmodern period, we expect the appearance of alternative professional types: the soldier–scholar, reminiscent of certain career officers in the period between the two world wars, and the soldier–statesman, the officer skilled in handling the media and adept in the intricacies of international diplomacy. Our argument is not that officers, having less to do in the way of war fighting, will turn to alternative pursuits. Rather effective performance of the officer's task in the postmodern period requires additional skills and capacities to justify on substantive grounds the military's role and its claim on social resources. While we can expect more than a residue of the warrior spirit to continue within the officer corps of the future, the relevant empirical question is which kind of officer will most likely be promoted into the military elite.

Finally, the civilian component of the defense establishment undergoes significant change as well. In modern military systems, civilians are a relatively minor component of the operational side of the defense establishment. But, in the late modern period, we see an increased number of civilian employees working in the defense establishment. In part this is due to the military's greater reliance on technically complex weapons systems, with the corresponding need for technical experts, both contract and directly hired, to work in the field.[22] There are, in addition, many nontechnical tasks, to include even menial labor which soldiers used to do, that have been turned over to civilian employment, usually on grounds of cost effectiveness.[23] In the postmodern period, we expect civilians to become even more intimately involved in military functions. This will occur in part as tendencies to move in this direction during the late modern period are reinforced. But more important, we think, will be increases in civilian–military involvement brought about through the distinctly postmodern merger of military roles with police, border, and security forces.

Consequences for Civil–Military Relations

With only some overstatement, we can describe the consequences of these changes for civil–military relations in the following terms. The mass army of the modern state was a unique institutional entity quite apart in its ethos and organization from civilian social structures. The late modern period, engaged in cold war, saw the rise of armed forces in which the skills of military personnel converged in many ways with civilian occupations. As compulsory military service systems weakened, defense establishments found themselves competing for personnel in the general labor market.[24] The distinction between the military and civilian society began to blur. This blurring launched an important and protracted scholarly debate about how unique the armed forces were (and should be) as institutions within contemporary society.[25] Nevertheless, the trend for the postmodern period points to a new model of the relation between the military and civilian society in which the distinction between the two is less significant. The military will be called upon to perform many nonmilitary functions while it retains a special responsibility for national defense. And the public will begin to recognize that both military and civilian service have some sort of civic equivalence.[26]

To trace the course of this movement we may begin by looking at the role of women in the military. Women were typically excluded from serving in the modern mass army, as they continue to be throughout most of continental Europe, especially in the Mediterranean countries. When they were allowed to serve, as they were in the United States and Britain, they were assigned to a separate corps. In the late modern period, women were partially integrated into the armed forces. Separate corps were abolished and quotas to limit the number of female enlistments were lifted. Yet women were excluded—sometimes by constitutional bars, sometimes by law, and sometimes by policy—from holding combat roles, a pattern that describes the position of military women in most European countries as well as in the United States at this moment. The situation, however, is in flux. In the postmodern military, pressure grows to incorporate women into all assignments, including combat roles. In the early 1990s steps in that direction, are being taken in the Netherlands and Canada and, in a more limited way, in Britain and in the United States as well. How far this movement should go has been a matter for intense debate in the aftermath of the Persian Gulf War.[27]

The questions raised have to do in part with the women's struggle for equality which is being carried on in all sectors of society. A modern mass army, more isolated from the larger society, might escape dealing with the effects of larger social movements. Civil–military relations, under that regime, are rather narrowly defined. The movement toward a postmodern army considerably widens the scope of concerns. The postmodern military is required to be much more responsive to changes in the larger society. We can observe the differences this makes, for instance, with respect to military families, toleration of homosexuals, and treatment of conscientious objectors.

In the old army of the modern period, a soldier's membership in the military community typically extended to his wife and family. The wife especially was expected to take part in numerous social functions and "volunteer" activities, and promotion to higher ranks might depend to some degree on how well one's wife performed in this role. In the late modern period, the wives and husbands (a new factor in the equation) of military personnel at both the noncommissioned and officer levels began to show reluctance to take part in customary military social functions.[28] In the postmodern military, with spouses of both sexes typically employed outside the home, we can expect fewer and fewer of them to have either the time or the inclination to engage in the social life of military installations. Importantly, their participation will not be expected by the military hierarchy. Curiously enough, it appears that as demands on military spouses decrease, there is more resentment at what demands remain.

The status of homosexuals in the military remains contentious, but the general movement is toward increased toleration. During the modern period in the mass army, military personnel who were discovered to be homosexuals were frequently incarcerated. Punishment was ineffective as a means of preventing homosexuals from performing military service. Codes of covert behavior coexisted with severe punishments.[29] In the late modern military of the Cold War era, homosexuals are still not welcomed, though the severity of punishment is diminished. On discovery, homosexuals in the military were simply discharged, with varying degrees of odium and the clear trend being toward the issuance of nonstigmatizing discharges. With increased dissensus in the postmodern period over the existence of certain, absolute standards to inform moral judgments, we expect that homosexuals in the postmodern military will be increasingly accepted, as presently occurs in the armed forces of Netherlands and, as of 1992, in Canada as well. In the United States, too, we should note that the higher education establishment for some years has urged the Defense Department to remove the bar on homosexuals entering the military as a condition for maintaining the Reserve Officer Training Corps (ROTC) on campus. Their concern is not just a matter of special interest pleading. The controversy surrounding President Bill Clinton's proposal, at first, to lift and, then, to ease application of the "gay ban" in the military shows how widespread public interest in this issue has become.

A common theme in all these debates about women, families, or homosexuals, is the conception of what the rights and duties of citizens are with regard to military service. In the modern period, military service was linked solely to the masculine role and conveyed formal rights to participate in the political life of the country. For men, the obligation to serve was compulsory. Deviance was not encouraged. When called upon, their family life was subordinated to the demands of the state. Under these circumstances, conscientious objection to military service was either prohibited, as was true of Mediterranean Europe and of the Warsaw Pact countries through the 1980s, or was confined mainly to members of traditional peace churches, as was true in the United States and Northern Europe through the 1960s. These attitudes depended in large part on high levels of popular support for the military establishment in its traditional form and acceptance of its critical role in maintaining national security. But as we move through the late modern period, and ambivalence toward military service grows, pressures mount on many fronts to loosen these restrictions. Women and homosexuals object to being excluded from military service, an exclusion which they see as limiting their rights as citizens to full participation in the political life of their countries and military families object to the demands of military etiquette which limit their participation in the larger their society.

Ultimately, there is less certainty that military and political authority have the right to compel young men to perform military service. The political forces pushing for an end to conscription are formidable and contradictory. They include traditional peace organizations, many religious groups, political radicals viscerally opposed to military establishments, libertarian conservatives who prefer to recruit the military on marketplace principles, government cabinet members looking for budget cuts, policy proponents who seek to transfer military spending to social programs, ethnic minorities who oppose being conscripted into larger state armies, young people imbued with individualism and materialism prevalent in Western youth culture, and even some military leaders who prefer to deal with a well-paid, well-trained professional force rather than with an army filled with reluctant draftees.

In this environment, the treatment of conscientious objectors has been liberalized and secularized considerably.[30] In the late modern period, with increased emphasis on a professional and technical military, conscientious objection becomes a widespread right, a trend clearly evident in North Europe. In addition, secular grounds for objection are gradually accorded the same standing in law as religious motives. And in the postmodern era, conscientious objection becomes a recognized right spreading to the former socialist countries of East Europe and the Mediterranean countries of NATO. In the postmodern society, we expect willingness to perform alternative civilian service to become the defining criteria of conscientious objection.

Yet we may expect even more change in this matter. The idea of alternative service presumes that systems of compulsory military service will continue in some form. But that is not a foregone conclusion. The trends of change in military organization which we have described in the context of an evolving postmodern society point rather in the opposite direction. Compulsory military service, which is still the dominant practice in all but the Anglophone countries, is unlikely to persist in its traditional form. If it fails to persist, we must face an unanticipated question. With high levels of opposition among citizens to military service, and the extraordinary proliferation in North Europe of alternative civilian service programs, what happens when conscription ends? Who will perform the vital social services now being undertaken by conscientious objectors avoiding military service? Here is a clear case where changes in military organization brought on by society return to affect the course of social change as well.

Obviously, one response to these questions is to establish a system of national service. In the United States, the debate on national youth service continues—indeed grows—despite the end of conscription. The idea of civic service programs for young people began in the early twentieth century with William James's famous formulation of "a moral equivalent of war." The Civilian Conservation Corps (CCC) of the New Deal era, the Peace Corps of the John F. Kennedy administration, and the Volunteers in Service to American (VISTA) were federal programs that garnered political and popular support. Since the 1970s, state and local voluntary youth service programs have burgeoned. In 1989, Senator Sam Nunn, the influential chairman of the Senate Armed Services Committee, introduced a bill that drew equivalencies between military and civilian service, and the current congress is dealing with President Clinton's similar, though scaled–down, proposal. Some of the most important precedents in the United States were the alternative programs developed for conscientious objection.

Perhaps no single pattern of relation between military and civilian service will emerge in a postmodern world. Yet, most likely, we can expect an end to any kind of evaluative differential between civilian and military service in those countries where conscription persists. Peace work will likely be accepted as legitimate service in countries further along on the postmodern road. We may even expect that those performing civilian service—and perhaps military service too—will be allowed to serve outside the boundaries of their home country.

The secularization and liberalization of conscientious objection, in sum, takes us on a postmodern turn. With the end of the Cold War, the service that used to be considered alternative becomes primary. The widespread military obligations of which citizens used to have are supplanted by new, broader forms of civilian service. In the postmodern society, military service itself becomes more androgynous and the citizen server replaces the citizen soldier.

Notes

1. Harold D. Lasswell, "The Garrison State," *American Journal of Sociology* 46 (January 1941): 455–468.
2. Surveys of the relevant literature are found, for example, in Peter Paret, *Understanding War* (Princeton, N.J.: Princeton University Press, 1992); Michael Howard, *War in European History* (New York: Oxford University Press, 1976); Theodore Ropp, *War in the Modern World*, rev. ed. (New York: Collier Books, 1962); J. F. C. Fuller, *The Conduct of War, 1789–1961* (New York: Da Capo Press, 1992); Morris Janowitz, *The Last–Half Century* (Chicago: University of Chicago Press, 1978), chap. 6.
3. See William Thompson's essay in chapter 3 of this volume.
4. For a theoretical discussion of the sources of this uneasiness, see David Sciulli's *Theory of Societal Constitutionalism* (Cambridge: Cambridge University Press, 1992), chap. 3.
5. The literature on postmodernism grows rapidly. Useful reviews for social scientists are John W. Murphy, "The Relevance of Postmodernism for Social Science," *Diogenes* 143 (Fall 1988): 93–110; David Harvey, *The Conditions of Postmodernity* (Cambridge, Mass.: Basil Blackwell, 1989); Anthony Giddens, *The Consequences of Modernity* (Stanford, Cal.: Stanford University Press, 1990); Scott Lash, *Sociology of Postmodernism* (New York: Routledge, 1990). Especially helpful for this essay was James Kurth, "The Post–Modern State," *National Interest*, 28 (Summer 1992): 26–35.
6. Jean–Francois Lyotard, *The Postmodern Condition*, trans. Geoff Bennington and Brian Massumi (Minneapolis: University of Minnesota Press, 1984), 81–82.
7. Roland Robertson has noted the sources of this phenomena in his essay, "After Nostalgia? Willful Nostalgia and the Phases of Globalization," in *Theories of Modernity and Postmodernity*, Bryan S. Turner, ed. (Newbury Park, Cal.: Sage, 1992): "One of the major features of modernity...is undoubtedly the *homogenizing* requirements of the modern nation–state in the face of ethnic and cultural diversity" (original emphasis). The triumph of nationalism, from 1750 to

1920, involved, therefore, "the attempt to overcome local ethnocultural diversity and to produce standardized citizens whose loyalties to the nation would be unchallenged by extra–local allegiances" (49).

8. Bryan S. Turner, "Periodization and Politics in the Postmodern," *Theories of Modernity and Postmodernity*, 11.

9. For a way of criticizing the quest for certainty that does not lead to nihilism, see Philip Selznick's *The Moral Commonwealth* (Berkeley and Los Angeles: University of California Press, 1992.)

10. See Kurth, "The Post–Modern State."

11. On these points see also Robert Reich, *The Work of Nations* (New York: Vantage, 1992).

12. Philip Wexler, "Citizenship in a Semiotic Society," *Theories of Modernity and Postmodernity*.

13. Joseph Schumpeter, *Capitalism, Socialism, and Democracy* (New York: Harper & Row, 1950).

14. Timothy Luke, "Televisual Democracy and the Politics of Democracy," *Telos* 70 (Winter 1986–87): 62, 72. Quoted by Wexler, "Citizenship in a Semiotic Society," 168.

15. Charles Tilly, *Coercion, Capital, and European States* (Cambridge, Mass: Basil Blackwell, 1990). See also Anthony Giddens, *The Nation–State and Violence* (Berkeley and Los Angeles: University of California Press, 1987) and Christopher Dandeker, *Surveillance, Power, and Modernity* (New York: St. Martin's Press, 1990).

16. Martin Van Creveld, *The Transformation of War* (New York: Free Press, 1991).

17. *New York Times* (August 21, 1991), A4.

18. For a comparative understanding of conscription systems, see James Burk, "The Decline of Mass Armed Forces and Compulsory Military Service," *Defense Analysis* 8 (Spring 1992): 45–59. A cross–national treatment of reserves systems can be found in Wallace Earl Walker, "Comparing Army Reserve Forces," *Armed Forces and Society* 18 (Spring 1992): 303–322.

19. For the American case see Murray Weidenbaum, *Small Wars, Big Defense: Paying for the Military after the Cold War* (New York: Oxford University Press, 1992).

20. Morris Janowitz, *The Professional Soldier* (New York: Free Press, 1960).

21. Charles C. Moskos, Jr. and Frank Wood, *The Military: More than Just a Job?* (New York: Pergamon, 1988).

22. Chris C. Demchak, *Military Organization, Complex Machines* (Ithaca, N.Y.: Cornell University Press, 1991).

23. There is too little research on the subject of civilians who work for the military. For one of the few analytical treatments, see Martin Binkin, *Shaping the Defense Civilian Work Force* (Washington, D.C.: Brookings, 1978).

24. The economic approach to manpower recruitment is documented in the President's Commission on the All–Volunteer Force (Gates Commission), *Report of the President's Commission on an All–Volunteer Force* (Washington, D.C.: Government Printing Office, 1970). See also William Bowman, Roger Little and G. Thomas Sicilla, *The All–Volunteer Force After a Decade* (New York: Pergamon–Brassey's, 1986).

25. Bernard Boëne, "How Unique Should the Military Be?" *European Journal of Sociology* 31 (1990): 3–59.

26. On the linkage between military and civilian national service see Morris Janowitz, *The Reconstruction of Patriotism* (Chicago: University of Chicago Press, 1983); Charles C. Moskos, Jr., *A Call to Civic Service* (New York: The Free Press,

1988); David R. Segal, *Recruiting for Uncle Sam* (Lawrence: University Press of Kansas, 1989).

27. The literature on this subject is vast and growing. For two recent and interesting overviews of the American case, see Mady Wechsler Segal and Amanda Faith Hansen, "Value Rationales in Policy Debates on Women in the Military: A Content Analysis of Congressional Testimony, 1941–1985," *Social Science Quarterly* 73 (June 1992): 296–309 and Jean Ebbert and Marie–Beth Hall, *Crossed Currents: Navy Women from WWI to Tailhook* (Washington, D.C.: Brassey's [US], 1993).

28. On changes in the military family, see Mady Wechsler Segal, "The Military and the Family as Greedy Institutions," in Moskos and Wood, *The Military*, 79–98.

29. For an account of homosexuals in the military during World War II, see Allan Berube, *Coming Out Under Fire* (New York: The Free Press, 1990).

30. This argument is developed more fully in Charles Moskos and John W. Chambers, eds., *The New Conscientious Objection* (New York: Oxford University Press, 1993).

7

Multinational Peacekeeping Operations: Background and Effectiveness[1]

David R. Segal and Robert J. Waldman

Peacekeeping and Peacekeepers: A Critical View

While multinational peacekeeping operations have become increasingly common in the 1990s, there is little agreement within the political, military, or academic communities about the participation of military personnel in peacekeeping missions nor about which nations should be involved as peacekeepers. This lack of consensus has been evident, for example, in debates among the Western European nations and the United States regarding appropriate peacekeeping participation in the remains of what was once Yugoslavia. Military forces still tend to regard their mission as the fighting of wars, and the containment of other nations' internal problems has been regarded as a diversion.

Neither is the concept of peacekeeping understood by the public. Public definitions of military operations are shaped extensively by the mass media, and peacekeeping, particularly when successful, has not been regarded as newsworthy. Thus, peacekeeping missions have been largely invisible, and the public has received little information about them.[2] However, peacekeeping missions, which the military has resisted and the media have ignored, are likely to become an increasingly important component of the operational repertoire of armed forces in the post–Cold War world, affirming an important three–decade old hypothesis in military sociology that has heretofore been largely ignored.

In his classic volume, *The Professional Soldier*, Janowitz argued that in an age of nuclear and other high–lethality weapons, the era of total war has ended.[3] He proposed that we think in terms of constabulary forces rather than military forces. Constabulary forces are characterized by their rapid deployability in the face of an international crisis and by the commitment of their leaders to use force with restraint, to preserve viable international relations rather than to win military victory.

Janowitz recognized that military men would oppose this redefinition of the military role, and more than two decades after introducing the constabulary concept into military analysis, he reported that "military personnel have rejected, or at least resisted, the concept of a constabulary because to them it sounds too much like police work."[4] Neither has the concept been embraced by the academic community. It has been resisted by people who more broadly have opposed the establishment of the study of peace as a scholarly field.[5] And it has been challenged by scholars who claim that the deployment of military forces on constabulary, or peacekeeping, missions does not contribute significantly to peace.[6]

The purpose of our analysis is to move beyond the current research literature, which consists primarily of case studies, and to explore conditions under which the deployment of military forces under United Nations's auspices does contribute to peace in conflict situations. Specifically we are interested in the impact of peacekeeping interventions on the control of conflict under varying conditions of self–deterrence among disputants: their willingness and inclination to seek peaceful resolution of their conflicts.[7] We assume that most parties, under most situations, would prefer peace to war.

The general proposition we seek to explore is that improving the legitimacy of peacekeeping interventions increases the likelihood of disputant acceptance—and so the success—of conflict control. We expect that as conditions of disputant self–deterrence deteriorate, the level of peacekeeping legitimacy becomes increasingly important in the ability of peacekeeping forces to control the conflict. Legitimacy, we think, rather than force accounts for peacekeeping success. Nevertheless, under extremely low conditions of disputant self–deterrence, no peacekeeping interventions are likely to be successful in promoting conflict control, and this is true no matter how legitimate the presence of peacekeeping forces is thought to be.

The collapse of the Warsaw Pact, and more importantly of the Soviet Union at the turn of the decade of the 1990s, ushered in a new era of multinational peacekeeping operations. This era, in contrast to earlier periods in peacekeeping history, is likely to be characterized by an increase in the legitimacy and number of such operations and their geographical scope, by their being conducted under the auspices of the United Nations rather than regional or other supranational bodies, and by major power military participation. In this paper we explore the evolution of multinational peacekeeping operations and the conditions under which such operations are likely to be effective. We begin with a discussion of the evolution of peacekeeping as a form of collective action undertaken by nations in their quest for peace and security. In the context of this history, we then present our empirical analysis of peacekeeping activities and their outcomes.

The Quest for Collective Security

The deployment of military personnel for peacekeeping, as opposed to war–fighting functions, can be seen as the latest stage in an ongoing attempt by the international community to resolve international differences without war, while at the same time providing for collective security. The nature of such peacekeeping deployments has changed markedly from the pre–Cold War to the post–Cold War period, as well as during the Cold War.

A century and a half before the Cold War, a series of Great Power multinational conferences was begun, and for almost a century, from 1815 to 1913, Prussia, Russia, Austria, and France attempted to act as a European Board of Directors, minimizing friction among nations.[8] More than thirty conferences were held, which did coordinate commerce and colonization, but failed to establish control over the military capabilities of the Great Powers. Recognizing this, Russia, the smaller European nations, and non–European nations met in The Hague in 1899 and 1907 to establish a peaceful settlement system for international disputes.

World War I demonstrated the failure of both conference programs, and the League of Nations emerged with the objective of preventing another war by default through broader international participation. The League, by combining the principles of a Great Power council, a general assembly of nations, and a functioning international administration, moved significantly toward the rationalization of conflict management, but did not achieve it. In its quest to achieve peace, the League attempted three pre–Cold War peacekeeping operations by military forces. It proposed to interpose a military force between Poland and Lithuania in 1921 but canceled its plans in response to a Russian protest; it fielded a (primarily Colombian) military force to mediate a dispute between Peru and Colombia over the Andean region of Leticia Trapeze in 1933; and it fielded the Saar International Force to oversee the plebiscite that merged that region with Germany in 1934–1935. Yet the League did not prevent World War II.

The end of World War II saw the emergence of the United Nations and the substitution of its principle of "peace through strength" for the League's principle of "peace through democracy." Between 1945 and 1970, during the early and mid–Cold War period, there were twelve United Nations peacekeeping missions in Greece (1947), Palestine (1949), Indonesia (1949), Kashmir (1949), Egypt (1956), Lebanon (1958), Congo (1960), West New Guinea (1962), Yemen (1962), Cyprus (1964), India and Pakistan (1965), and the Suez Canal (1967). However, the Cold War proved to be a constraint on the effectiveness of military forces to contain or control, rather than to wage, conflict.

Larry L. Fabian saw these twelve operations as spanning three generations, from 1944 to 1970, of United Nations Cold War peacekeeping preparedness.[9] The first generation, 1944–1955, was characterized by overt hostility between the Soviet Union and the United Nations, rooted in part in the domination of the United Nations by the West and in part by Stalin's leadership style. There was no clear international constituency for United Nations peacekeeping preparedness, and the level of such preparedness was low. The year 1953 set the stage for the advent of the second generation; Josef Stalin died, and Dag Hammerskjold was elected U.N. Secretary General. The second generation, 1956–1965, was the only one prior to the current period to begin with any promise for peacekeeping preparedness. The new Soviet leadership was more flexible, the Cold War was less intense, and Hammerskjold built a middle–power constituency for peacekeeping preparedness from the expanding membership of the United Nations. In addition, he enjoyed the active support of the United States, and initially, the passive acquiescence of the Soviet Union, a resource his predecessor, Trygve Lie, had lacked.

However, Soviet opposition to the peacekeeping operation in the Congo, the largest and costliest of the international body's peacekeeping operations, shattered consensus. By 1966, Soviet opposition to a Canadian peacekeeping proposal immobilized the secretary general and his middle–power constituency. The third generation of peacekeeping preparedness, from 1966 to 1970, was distinguished by recognition of the importance of superpower approval for any plan to build a true international peacekeeping force. In part a reversion to the first generation, this stage was characterized by unilateralism, a lack of formal structure, diffusion, and experimentation.

On the basis of these operations, Fabian developed a program for the establishment of a United Nations peacekeeping capability which focused largely on the relationship between the United States and the Soviet Union. The program required that the United States and the Soviet Union reach consensus on issues regarding the exercise of broad political responsibility for peacekeeping preparedness, that the superpowers share this responsibility, and that their role in actual peacekeeping activities be minimized, with the states actually involved in peacekeeping being broadly representative of U.N. members, including third world and Eastern European states.

Relations between the superpowers in the 1970s and 1980s did not meet the conditions posed by Fabian, and peacekeeping during this period continued to be largely unilateral, diffuse, informal, and experimental. Indeed, the 1967–1973 period was one of U.N. peacekeeping dormancy, with no new operations authorized or implemented,[10] and during the "resurgent period" of the mid– and late 1970s, only three operations were mounted, all in response to the crisis in the Middle East, which has become the crucible of peacekeeping: the United Nations Emergency Force II (UNEF) on the Israeli–Egyptian border; the United Nations Disengagement Observer Force (UNDOF) on the Israeli–Syrian border; and the U.N. Interim Force in Lebanon (UNIFIL). During the early 1980s, reflecting the Cold War strains between the United States and the Soviet Union, two additional multinational peacekeeping operations were mounted in the Middle East under auspices other than the United Nations, viz., the Multi–National Force (MNF) in Lebanon, manned by troops from four NATO nations, and the Multinational Force and Observers (MFO) in the Sinai. This last operation has drawn on personnel from twelve nations who report to a newly invented civilian directorate.

Reflecting concerns in the United States about a potentially expanded Soviet influence in the Western Hemisphere, United Nations peacekeeping was largely restricted to Africa and the Middle East during the Cold War. In the Western Hemisphere, the United States has supported the concept of multinational peacekeeping, but it has been carried out under regional rather than global auspices. Reliance on the United Nations has been avoided. During the 1965–1966 crisis in the Dominican Republic, for example, the United Nations established a Mission of the Representative of the Secretary General in the Dominican Republic (DOMREP), but its total strength was only two military observers. The

U.S. provided most of the personnel for the peacekeeping force which was established under the auspices of the Organization of American States (OAS) and effectively controlled the situation. Similarly, almost two decades later, the American invasion of the Caribbean island of Grenada was ostensibly encouraged by the Organization of Eastern Caribbean States. A joint Caribbean task force of 300 soldiers and police from Antigua, St. Lucia, St. Vincent, Dominica, St. Kitts, and Barbados followed the U.S. troops to the island and, after the U.S. withdrawal, a Caribbean Peacekeeping Force remained in place.

The democratic revolution in Eastern Europe in the late 1980s might itself have set the stage for the achievement of Fabian's conditions and ushered in a new generation of United Nations peacekeeping efforts as he envisaged them under consensual superpower support, had the Soviet Union survived the upheaval. This possibility was reflected, for example, in agreement between the U.S. and the USSR, after considerable negotiation, on U.N. Security Council action responding to Iraq's 1990 invasion of Kuwait. However, the deconstruction of the Soviet Union after the attempted August 1991 coup made superpower relations—and Fabian's blueprint for peacekeeping success—irrelevant factors. Most impressively, in the wake of the collapse of the Soviet Union, no fewer than six new U.N. peacekeeping missions were initiated or planned in the 1991–92 period alone: half as many as were initiated in the first quarter of a century of United Nations peacekeeping operations, between 1945 and 1970.

These new missions are geographically diverse, spanning Southwest and Southeast Asia, Africa, Latin America, and Central Europe. These are the U.N. Iraq–Kuwait Observation Mission (UNICOM), the U.N. Angola Verification Mission (UNAVEM II), the U.N. Observer Mission in El Salvador (UNOSAL), the U.N. Mission for the Referendum in Western Sahara (MINURSO), the U.N. Transitional Authority in Cambodia (UNTAC), and the U.N. Protection Force in Croatia (UNPROFOR).

These initiatives suggest that the legitimacy of multinational peacekeeping under United Nations auspices has increased. Most of these missions have or are slated to have major power participation. A new generation of U.N. peacekeeping efforts is indeed unfolding. Unlike the previous generations, it is not being shaped or constrained by superpower relationships.

The use of military personnel from the major powers or superpowers in peacekeeping forces has nonetheless remained a debatable issue in peacekeeping doctrine. Charles C. Moskos, in his research on the

U.N. mission in Cyprus, had hypothesized that "Soldiers from middle powers are more likely to subscribe to the constabulary ethic [a commitment to the avoidance of use of force and to the maintenance of viable international relations] than are soldiers from major powers."[11] His findings led him to soften his position. He found no difference in constabulary orientation between British soldiers assigned to UNFICYP and soldiers from neutral middle powers. Nonetheless, he concluded that the realities of international politics were also critical elements to the success of peacekeeping missions, making the neutral middle powers the primary and appropriate source for peacekeeping forces.

In a bipolar world, it was difficult to regard either of the superpowers, or their major allies, as impartial parties anywhere in the world. Moreover, as history has shown, superpower military forces on peacekeeping duty are attractive and vulnerable targets for terrorist attack. Witness the October 1983 bombing of the Marine headquarters in Beirut,[12] and the recent allegations of terrorist responsibility for the 1985 crash in Gander, Newfoundland, of a plane load of American soldiers from the 101st Airborne Division returning home from peacekeeping duty in the Sinai MFO. Moskos's conceptualization, like Fabian's, was rooted in the geopolitical realities of the bipolar Cold War world. The collapse of the Soviet Union renders some of their assumptions moot. And the experience of the early 1990s, particularly on the Arabian Peninsula and in the former Yugoslavia, suggests that in fact the effective functioning of multinational military forces might require superpower or major power participation. While their past doctrinal exclusion precludes empirical evaluation of the implications of their participation, it is possible to evaluate past U.N. peacekeeping operations to determine lessons that might be useful in the future, as peacekeeping by military forces becomes increasingly common. It is to this task that we now turn.

Data Base

In order to develop an empirical data base on past multinational peacekeeping activities and their consequences, prior peacekeeping missions were screened against three criteria. First, we sought missions that were generally regarded as reflective of three central tenets of contemporary peacekeeping doctrine: impartiality, consent of disputant parties, and noncoercion, to the greatest extent possible. This criterion excluded early formative missions in Greece and Indonesia.

Second, we focused on peacekeeping interventions conducted under the auspices of the United Nations rather than regional or ad hoc organizations. On the one hand, we felt that conditions of peacekeeping efficiency might vary as a function of sponsoring organization. But we did not have enough cases of peacekeeping under the auspices of organizations besides the United Nations which would allow us empirically to analyze this variation. On the other hand, we anticipate that with the demise of the Soviet Union, future peacekeeping missions are increasingly likely to be United Nations operations, and, if that is so, U.N. missions are the most useful source for lessons applicable in the future. Applying this criterion excluded interventions such as the Sinai MFO from our data base.

Finally, we excluded interventions initiated after 1988, because of our dependence on documentary sources of data and the limited information available on recent peacekeeping missions.

Eleven peacekeeping missions, with fairly broad geographical range, met these three criteria. These were: UNFICYP (Cyprus) in the West; UNOC (Congo/Zaire) in Africa; UNMOGIP and UNIPOM (India/Pakistan) and UNTEA (Indonesia) in Asia; UNTSO, UNEF I and UNEF II (Sinai), UNYOM (Yemen), UNDOF (Golan Heights), and UNIFIL (Lebanon) in the Middle East.

Across these eleven missions, we identified from the peacekeeping literature and from *The News Dictionary* a purposive sample of 100 specific peacekeeping events. A peacekeeping event was defined as a specific confrontation in the course of a peacekeeping mission in which two opposing disputants, a peacekeeping intervention, and a conflict outcome could be identified through documentary evidence. Thus, while peacekeeping missions are our objects of analysis, specific peacekeeping events that take place in the course of these missions are our units of analysis. We believe that peacekeeping missions have their own natural histories, and that the overall long–term success or failure of a peacekeeping mission is largely determined by the ability of peacekeepers to contain specific conflict events as they unfold.

In order to be assured of variation in outcomes, and having no a priori knowledge of the ratio of success to failure in peacekeeping events, we purposively selected 50 events in which the peacekeeping initiatives were accepted by the parties to the conflict and 50 in which the initiatives were rejected. We initially wondered whether we could in fact document 100 events. Very quickly, however, the problem became one of selecting only 100 events from an expanding complex of docu-

TABLE 7.1
Conflict Control Acceptance and Rejection by Mission

Mission	Number Accepted	Number Rejected
UNFICYP	10	11
UNOC	11	11
UNMOGIP–UNIPOM	7	8
UNTEA	3	0
UNYOM	1	2
Israeli Border	18	18
Total	50	50

mented intervention situations, since the natural history of a peace-keeping mission is composed of a series of engagements and disengagements. In making these selections, we focused on events which identified key turning points in the peacekeeping missions, and which reflected the variety of experiences in each mission.

Most of the events took place on the Israeli border, in Cyprus, or in the Congo. As table 7.1 shows, the distribution of events across U.N. missions was: UNFICYP, 21; UNOC, 22; UNMOGIP–UNIPOM, 15; UNTEA, 3; UNYOM, 3; Israeli Border Missions, 36. Examples of the events include the supervision of a cease–fire agreement between India and Pakistan by UNMOGIP in January 1949, the Israeli invasion of Egypt in October 1956, a Greek Cypriot attack on St. Hilarion Castle in April 1964, the invasion of Kasai by Congolese troops in August 1960, a clash between Papuan police and Indonesian soldiers in December 1962, and a terrorist attack on an Israeli village in August 1981.[13] As table 7.1 reflects, the distributions of events in which conflict control was accepted, and where it was rejected, were held roughly equivalent across peacekeeping missions.

Having selected these events, we then established a data file based upon available historical records. Drawing on historical documentation, the peacekeeping interventions were coded for the level of force used by peacekeepers in each event. The use of peacekeepers for observation and patrol anchored the low end of this variable, with interposition, riot control and sanctions in the middle of the range, and the aggressive use of force at the high end of the scale. As we would expect, the distribution of events was concentrated at the low–force end. In the plurality of events (42) the peacekeeping intervention was routine observation or patrol, and in another 17 cases it involved local mediation or public service. In 18 cases, intervention was used as a bargaining chip in higher level negotiations. In 7 cases, the peacekeepers elected not to use force in response to attack. In 8 cases, force or sanctions were threatened but not used. At the high end of the scale, the peacekeepers used force aggressively in 6 events, for example, attacks by U.N. forces on Katangan troops, and defensively in 2 events.

Peacekeeping outcomes were classified as accepted (and if so, the nature of the cease–fire achieved), or rejected (and if so, the nature of the continuing violence). For example, the 1964 negotiation of a cease–fire after extensive Turkish bombings in northeastern Cyprus was defined as an acceptance of or successful conflict control, while riot control interventions in the Congo that did not result in an end to the rioting were defined as rejections or failures of conflict control. The coding of interventions and outcomes involved factual data that could be verified from two or more sources.

For each conflict event we also collected data on conditions of disputant self–deterrence, based upon the recent history of hostilities prior to the event, and on perceptions of vulnerabilities. This dimension ranged from cases where violence was regarded as *improbable* (11 events) because in the unfolding of events prior violence had stopped and discussions had been initiated and/or disputants were committed to trying nonmilitary options, through those where violence was regarded as *uncertain* (38 events) due to the presence of volatile and heavily armed disputants or to recurring violence under conditions that provided salient benefits for ending violence, to cases where violence was regarded as *probable* (23 events) due to recent violence and continued mobilization for violence, and those where violence was regarded as *certain* (28 cases) because one disputant would clearly profit from a violent solution, as in India in 1971 after the revolution in East Pakistan, or because of a total breakdown in law and order, as in Southern Lebanon. An anticipated fifth theoretical category, violence *impossible*, was dropped from the analysis because the historical data on our 100 events provided no such empirical cases.

The legitimacy of the peacekeeping intervention (commitment of disputants to the process; incentives or disincentives for compliance) was coded on a scale ranging from low to high support for peacekeeping. At the low end of the scale (29 events), mission authorization or consent had been disputed or withdrawn, but some peacekeeping operations continued. (Events that occurred after peacekeeping forces withdrew after a withdrawal of consent were not included in our analysis.) An additional 32 events were regarded as having indifferent or ambivalent support because of mandate disagreements or failures to cooperate despite the existence of formal consent. In these cases, peacekeeping operations were limited to routine patrol operations. Another 15 cases were classified as having contingent support. In these cases, disputants had formally consented to the intervention, had disputed U.N. mandate understandings or failed to comply with mandate requirements, but the peacekeeping mission was actively involved in conflict resolution efforts. At the high legitimacy end of the scale (25 events), disputants had formally consented to peacekeeping intervention, accepted the intervention mandate as defined by the U.N., and the peacekeeping mission was actively and directly involved in conflict resolution efforts (mediation, local public service, use of force).

Because the self–deterrence and legitimacy dimensions involve more subjective judgment on the part of coders than did other variables, an extensive reliability analysis of the coding was undertaken. Seven coders with expertise in analyzing international events independently coded ten events. We were initially concerned by what appeared to be a low level of inter–rater reliability in our measure of self–deterrence. However, analysis revealed that the problems involved a small number of events on which historical data was limited, and disagreement over contiguous code categories. A more precise definition of events and collapsing of ambiguous categories produced a 94 percent level of coder agreement. Similarly, a redefinition of categories in the legitimacy scale produced a 90 percent level of coder agreement.

The disputant pairs in each event were also coded to indicate whether they were both represented by conventional military forces (e.g., the Greek vs. Turkish armies in the Cyprus case), neither was represented by conventional military forces (e.g., Greek Cypriots vs. Turkish Cypriots), or one was represented by a conventional military force and the other not (e.g., Turkish army vs. Greek Cypriots). The distinction among the three categories of disputant pairs was based upon the assumptions that integral actors are more likely to resolve conflicts cooperatively than are factionalized actors,[14] and that conventional military forces are more likely to be integral actors (i.e., disciplined) than are unconventional military forces. In 64 of the events, the disputant pairs were both represented by conventional military forces. In 17 cases, only one had a conventional military force, and in 19 cases, neither had a conventional force.

Contrary to our expectations, however, the nature of the military organization of the disputants did not affect the overall outcome of peacekeeping interventions, although it had a slight relationship to the efficacy of force as a peacekeeping intervention. The general lack of relationship to peacekeeping outcomes is itself an interesting and counter–intuitive finding.[15]

Analysis

As anticipated, we found the likelihood of peacekeeping success to vary inversely with the level of disputant self–deterrence (estimated probability of violence in the conflict situation). As table 7.2 shows, peacekeeping initiatives were successful in all 11 events in which non-violence was regarded as probable. Indeed, this finding is almost tauto-logical, and says little other than that our coding of nonviolence probability was highly reliable. However, peacekeeping was also suc-cessful in almost two–thirds of the events in which violence was re-garded as uncertain. Perhaps even more impressive, peacekeeping was successful in more than one–quarter of the cases in which conditions seemed to make violence probable, and in almost one–third of the cases in which violence seemed certain. While the data suggest that our esti-mates of the likelihood of violence were reliable all across the self–deterrence spectrum, they also suggest that peacekeeping efforts can sometimes deter violence even when it is a highly likely outcome. We now consider the conditions under which this success was achieved.

Certain Violence

As noted earlier, we had anticipated that at lower levels of disputant self–deterrence, peacekeeping legitimacy would become increasingly important in conflict control, that legitimacy rather than force would ac-count for peacekeeping success, and that at low levels of disputant self–deterrence no peacekeeping interventions were likely to be successful. Table 7.2 shows this last expectation to have been overly pessimistic.

Our analysis revealed for those cases in which we regarded violence a certainty, the first two expectations were not wholly correct either. Among the 28 events in which violence was regarded as certain, legiti-macy was low in most (23) cases, but there was a significantly greater likelihood of conflict control success in the few cases where legiti-macy was high (4 cases out of 5) than when it was low (5 cases out of

TABLE 7.2
Peacekeeping Event Success and Failure under Varying
Conditions of Disputant Self–Deterrence

Control of Conflict	Violence Certain	Violence Probable	Violence Uncertain	Nonviolence Probable
Successful	9	6	24	11
	(32.1%)	(26.1%)	(63.2%)	(100%)
Failure	19	17	14	0
	(67.9%)	(73.9%)	(36.8%)	(0%)
Total	28	23	38	11

23). Nonetheless, there were 5 cases of successful peacekeeping under conditions of violence certainty and low peacekeeping legitimacy. Of the nine successful cases of peacekeeping, six of them involved peacekeeping interventions that required the use of force or threat of force, rather than being restricted to mediation or patrols. By contrast, ineffective interventions under conditions of violence certainty primarily involved patrols. Success at conflict control was more strongly associated with level of force (lambda=.444) than with legitimacy (lambda=.333) under these conditions. (We use lambda as a nonparametric measure of the strength of association when comparisons of effects are made. Our small case base precludes higher order multivariate analyses.) Table 7.3 presents these bivariate relationships.

Probable Violence

Our expectations received more support among those events in which we had regarded violence as probable but not certain. As table 7.2

TABLE 7.3
Impact of Level of Legitimacy and Level of Force on Sucess of
Conflict Control under Conditions of Certain Violence

Conflict Control	Level of Legitimacy		Level of Force		
	Strong or Conditional	Ambivalent or Critical	Patrol	Mediation	Force
Successful	4	5	1	2	6
Not Successful	1	18	15	2	2
lambda	.333		.444		

TABLE 7.4
Impact of Level of Legitimacy and Level of Force on Sucess of
Conflict Control under Conditions of Probable Violence

Conflict Control	Level of Legitimacy		Level of Force		
	Strong or Conditional	Ambivalent or Critical	Patrol	Mediation	Force
Successful	5	1	2	4	0
Not Successful	0	17	14	2	1
lambda	.833		.333		

showed, peacekeeping success was somewhat lower under conditions of violence probability than violence certainty.

Only 6 of 23 events studied in this category resulted in the acceptance of conflict control. However, 5 of these 6 events were characterized by high peacekeeping legitimacy or at least conditioned support, and no high legitimacy or conditioned support interventions failed when legitimacy was high, even though conditions led to a judgment that violence was probable. Moreover, mediation rather than the use of force was the most effective intervention strategy. (Indeed, of the 6 events in this category, 4 entailed mediation and two involved patrols; none involved use of force). The mediation generally involved some action by senior U.N. or peacekeeping force individuals. In these cases, peacekeeping legitimacy was strongly associated with conflict control success (lambda=.833), and the relationship between use of force and peacekeeping success was considerably weaker. These data are presented in table 7.4.

TABLE 7.5
Impact of Level of Legitimacy and Level of Force on Sucess of
Conflict Control under Conditions of Uncertain Violence

Conflict Control	Level of Legitimacy		Level of Force		
	Strong or Conditional	Ambivalent or Critical	Patrol	Mediation	Force
Successful	18	6	3	15	6
Not Successful	1	13	10	3	1
lambda	.786		.500		

Uncertain Violence

Nearly half of the successful interventions in the events that we studied (24 of 50) occurred under violence uncertainty conditions. Success at conflict control was strongly related to peacekeeping legitimacy under these conditions as well. Of 19 events in which legitimacy was strong or conditional, 18 resulted in successful interventions. However, of 19 events where support for peacekeeping by the disputants was ambivalent or critical, only 6 resulted in successful conflict control. The association between legitimacy and peacekeeping success (lambda=.786) was higher than that between use of force and success (lambda=.500). Mediation was the most common form of intervention under these conditions, and it was successful in 15 out of 18 instances. In proportional terms, the use of force was as effective as mediation, but it was much less commonly applied. Mere presence—the use of peacekeeping patrols—was largely ineffective as a means of eliciting compliance. This was the intervention used in 10 of the 14 unsuccessful peacekeeping events under conditions of violence uncertainty. Patrolling had been a relatively ineffective peacekeeping tool under conditions of violence certainty and violence probability as well.

Conclusions

In an era characterized by increasing peacekeeping activity under U.N. auspices around the globe, it is useful to consider the likelihood of success of such missions by putting the assessment in historical perspective. Peacekeeping interventions have not been likely to stop violence initiated by a disputant moving toward a military victory after negotiations have failed. Peacekeeping interventions also have had a low likelihood of success when disputants have been mobilized for ongoing or imminent violence, or when disputants, with growing hostility and enhanced arms, have been moving toward mobilization for imminent violence or a spiral of violence driven by a quest for revenge even without hope of victory. These situations have two characteristics in common. First, the disputants confront potential losses, including opportunity costs and increased risks, if they do not use violence. Second, the disputants do not see viable nonviolent options. Even under these conditions, however, peacekeeping has achieved occasional success.

Peacekeeping interventions have been most likely to succeed when disputants have experienced the costs of violence and, in addition, have at least a potential stake in pursuing nonviolent options. Peacekeeping interventions have also been most likely to effect conflict control when both disputants have something to gain from peacekeeping's success.

Peacekeeping interventions have been more likely to succeed if they have been opportunistic, being proactive in establishing a linkage between compliance and situational advantage for both disputants. Opportunistic interventions have been the most likely to succeed under conditions in which law and order have broken down, violence has been probable, and hostility has been spiraling upward. And they have been successful even when peacekeeping legitimacy was low, although legitimacy clearly has an effect on peacekeeping success. Strong formal consent only emerged for opportunistic interventions, and under conditions of relatively low likelihood of violence (violence uncertainty and nonviolence probability), strong consent has only emerged after situational advantages have been established.

Force or the threat of force as well as mediation have proven to be important factors in creating opportunistic interventions. In a sense, while peacekeeping in idealistic terms is not a job for soldiers, pragmatically only soldiers can do it, for routine patrolling may develop into the application of force. And in the absence of bipolar confrontation in the post–Cold War world, superpower or major power participation in peacekeeping operations may increase the credibility of the threat of force, with the potential interrelated second–order consequences of reducing the likelihood that force will have to be used, and increasing the likelihood that if force is used, the peacekeepers will prevail. However, there are important limitations to the use of force as a peacekeeping tool. It has been effective primarily in local situations involving disputants that are not conventional military units. And it is likely to be regarded as impartial and legitimate only where the use of force does not disadvantage any consenting disputant.

It is difficult to demonstrate a direct causal link between peacekeeping interventions and peacekeeping outcomes given the complexity of conflict situations and the relatively small number of events with which we are working in this exploratory analysis. However, the association between proactive mediation or threat or use of force and conflict control is clear, and conflict has historically been prevented or diminished in circumstances that, absent the intervention, would have been expected to have been more violent. This study has also demonstrated the feasibility of establishing a data base for the analysis of peacekeeping operations across missions and over time. Expansion of such a data base would facilitate modeling and analysis to help understand the complexity of these relationships.

Notes

1. This paper was written during the first author's tenure as the S. L. A. Marshall Chair at the Army Research Institute. The interpretations in this paper are those of the authors, and not necessarily of the Army Research Institute, the Department of the Army, or the Department of Defense. We are indebted to James Burk, Ted Robert Gurr, Barbara F. Meeker, Jerald Hage, and Joseph Lengermann for their comments.
2. David R. Segal, Mady Weschler Segal, and Dana P. Eyre, "The Social Construction of Peacekeeping in America," *Sociological Forum* 7 (1992): 121–136.
3. Morris Janowitz, *The Professional Soldier* (New York: The Free Press, 1960).
4. Morris Janowitz, "Civic Consciousness and Military Performance," *The Political Education of Soldiers*, Morris Janowitz and Stephen D. Wesbrook, eds. (Beverly Hills, Cal.: Sage, 1983), 55–80.
5. See, for example, Donald Kagan, "The Pseudo–Science of 'Peace,'" *Public Interest* 78 (Winter 1985): 43–61.
6. Paul F. Diehl, "When Peacekeeping Does Not Lead to Peace," *Bulletin of Peace Proposals* 18 (1987): 47–53.
7. Robert J. Waldman, *International Peacekeeping: Conditions of Conflict Control* (Ph.D. dissertation, University of Maryland, College Park, 1991).
8. Inis L. Claude, Jr., *Swords into Plowshares* (New York: Random House, 1984).
9. Larry L. Fabian, *Soldiers without Enemies* (Washington, D.C.: Brookings Institution, 1971).
10. Henry Wiseman, *Peacekeeping: Appraisals and Proposals* (New York: Pergamon, 1983).
11. Charles C. Moskos, *Peace Soldiers* (Chicago: University of Chicago Press, 1976), 7.
12. David R. Segal and Katherine Swift Gravino, "Peacekeeping as a Military Mission," *The Hundred Percent Challenge*, Charles Duryea Smith, ed. (Cabin John, Md.: Seven Locks Press, 1985), 64.
13. A complete listing of these events can be found in Robert J Waldman, *International Peacekeeping*.
14. Christopher W. Moore, *The Mediation Process* (San Francisco, Cal.: Jossey–Bass, 1987).
15. A full description of the coding scheme as well as a copy of the data base appear in appendices in Robert J. Waldman, *International Peacekeeping*.

8

The Challenge of Nonviolence
in the New World Order

Robert L. Holmes

History may well record the social and political changes of the twentieth century's closing years as unrivaled in swiftness, scope, and consequence. Time has, to be sure, always rearranged the landscape of human affairs in its course. But wars and revolutions notwithstanding, this has often occurred gradually, one era's promontories of power metamorphosing almost imperceptibly into the next era's valleys of decline. Abrupter changes, even when they were to be of momentous eventual consequence, were localized in immediate effect because of limitations in travel and communication.

That has now changed. With benefit of near–instantaneous communications, today's major events reverberate quickly throughout the world, their effects augmented by the increased capacity for global action and reaction. Most dramatic of such recent events has been the transformation of Eastern Europe and the Soviet Union. In little more than a moment by historical standards, a great nation, whose rivalry with another great nation shaped the international scene for nearly fifty years, has ceased to exist, and along with it an empire forged in the name of a global ideology. This has left America as the world's sole superpower, with the problem of what to do with its vast military power. Its response to that challenge will, more than the actions of any other nation, largely determine the direction of humankind's course and the nature of the emerging new world order, at least for the foreseeable future. As put disarmingly by General Colin L. Powell, Chairman of the Joint Chiefs of Staff, "We no longer have the luxury of having a threat to plan for. What we plan for is that we're a superpower."[1]

What planning to be a superpower means needs translating, of course, and there is growing debate between those, on the one hand, who want the U.S. to rein in its military power to focus upon domestic needs and those, on the other hand, who relish the prospect of the virtually unimpeded exercise of that power abroad. Neo–isolationism and neo–imperialism, as these positions may be called respectively, represent extremes on a spectrum that admits of many gradations.

Neo–isolationism and neo–imperialism are normative positions, prescribing how the U.S. ought to proceed in the years ahead. It is with the normative dimension of the world situation that I am primarily concerned. Description and explanation are useful only if brought to bear on the guidance of conduct. However much some may deny it, virtually all analysts of the international scene implicitly or explicitly advance normative claims. And all of them, without exception, rely upon normative assumptions, if only because it is impossible to do otherwise.

But normative judgments require understanding the world. And different observers looking at the same facts following the collapse of the Soviet Union perceive different things. Some see a unipolar world with the U.S. as the dominant power. Others see a multicentric concert of powers, consisting of the U.S., Europe, Russia, Japan, and China.[2] Still others see an essentially intact but evolving state–centric system in which national sovereignty still has preeminence.[3] And still others question whether there is a new world order to be seen at all.[4]

Two Conceptions of a New World Order

Let me begin by saying what the new world order is not. It is not the order of the Bush administration's Gulf War rhetoric, according to which strong U.N. action in the Gulf crisis has ushered in an era of peace and order through collective security. The U.N. actions from August 8, 1990 on were largely orchestrated by Washington, which sought a military solution to the crisis virtually from the outset. More importantly, Washington made clear—most emphatically from its August 12th announcement of a unilateral blockade of Iraqi and Kuwaiti shipping—that it did not need Security Council authorization (or congressional either, for that matter), for its actions. Only in the face of growing criticism within the U.N. did it undertake an intensive diplomatic campaign to get that authorization, which it succeeded in doing on August 25th. Far from representing the realization of the heralded role for which the U.N. was created, these events transformed the Security Council almost overnight from a position of ineffectuality brought about by years of super-power neglect and obstruction to one of near–subservience to the remaining superpower.

So if by a new world order is meant one fashioned deliberately by the Bush administration and heralding a new era of international peace and order, there is no such thing—scarcely, any longer, even in the rhetoric of the U.S. government. But if by a new world order one means simply a radically changed world, with realigned powers and new challenges, then a new world order there will inevitably be. For when the transitional period we are now in stabilizes, there will have taken shape a world that, at least in traditional geopolitical terms, cannot help but be radically different from the one that has existed since the end of the Second World War.

A major candidate for such an order is set forth in the Pentagon Defense Planning Guidance for the 1994–1999 fiscal years.[5] The role initially envisioned there for the U.S. was that of preventing other nations from ascending to superpower status and maintaining a world order congenial to U.S. interests. Although that goal appears to have been dropped from a later draft, more moderate in tone, the document clearly sees the U.S. as the world's paramount military power and conducting itself accordingly.[6] Conflict in third world countries has been thought for some time to be the likeliest military challenge to the U.S. With Grenada, Panama, and the Gulf War as models, the expectation is that the U.S. military will visit swift, massive, and decisive force upon those countries, or elements therein, who seriously threaten that status or the order on which it depends. To the extent that this represents the de facto position of the U.S. government, which remains to be seen, it brings policy planners down squarely on the side of neo–imperialists.

Thus two conceptions of a new world order need to be reckoned with. The first, which many believe to have been born of the Gulf crisis, and which I shall call WO1, envisions collective security and international law as the paramount features of the international world. It represents a refurbishing of the conception George Kennan saw the U.S. aspiring to in much of the first half of the twentieth century.[7] The second, which I shall call WO2, may maintain the appearance of WO1, including some of WO1's structures (such as the International Court of Justice and the U.N.). But it envisions the U.S. as remaining the sole global superpower, whose will and determination guarantee a stability designed to enable its interests to flourish. It represents the world dreamed of by political realists, in which power and national interest, rather than legalistic/moralistic norms, prevail; a world transfigured by the collapse of the Soviet Union to feature a resplendently victorious United States.[8] WO1 is internationalist in character, but allows wide latitude in how extensive nations' international commitments will be. It is compatible with all but the more extreme forms of neo–isolationism. But it is incompatible with neo–imperialism, which, though it is internationalist in character, is individualist rather than collectivist in its conception of the form such internationalism should take. WO2, on the other hand, represents an imperial order, consistent only with, indeed embodying the objective of, the neo–imperialist philosophy.

Will Either World Order Prevail?

Whether either WO1 or WO2 will in fact be realized in anything like full–blown form remains to be seen. But three factors will almost certainly play a major role in determining the answer to that question.

The first is that even before the close of the century, the defeated nations of the world's greatest war loom once again as major powers. Japan is not yet a military power, but despite a constitutional ban on remilitarization there is a growing aspiration among many on Japan's right to restore Japanese military might—an aspiration encouraged by the U.S. throughout much of the Cold War, as it sought to counteract Soviet influence in Asia. The historic vote of the Japanese parliament to allow deployment of Japanese troops abroad, even as only part of a U.N. peacekeeping force, is an ominous step in that direction. Germany is a military power, however, though with fragile constraints against the use of that force outside of NATO, and is now looking for a new mission for its military in the post–Cold War era.[9] Some are calling for "selective" nuclear proliferation to allow Germany to become a nuclear power as well.[10] Meanwhile, neo–Nazism—a resurgence of the very plague that led the world into World War II—is alive and flourishing in Germany, various Eastern European countries, and the U.S., and proto–Nazi nationalism asserts itself in Russia and what remains of Yugoslavia.

The second is that as the West glows with satisfaction over the collapse of communism in the Soviet Union, a powerful, revolutionary Marxist movement, the Shining Path, with the potential of creating another Vietnam, gains strength in Peru; and another, the Khmer Rouge, even more ominous because of its past record, becomes a growing threat in Cambodia. Their rhetoric aside, the ruthless, oppressive, centralized bureaucracies of the Soviet Empire bore about as much resemblance to the Marxist vision as Western nations do to Christ's vision in the Sermon on the Mount. It would be naive to suppose that the collapse of that empire signals the end of Marxism's appeal to the world's disadvantaged, particularly to those in the third world who feel the continuing effects colonialism's legacy.

The third, and perhaps least appreciated, is the extent to which Marxist predictions for capitalism show signs of being confirmed by developments in the U.S.: increasing numbers (now more than 30 million) living in poverty, a shrinking middle class, increasing concentration of wealth in the hands of a few, and—conforming to Lenin's projection for the final, imperialistic stage of capitalism—a flight of American companies, to the third world in search of cheap labor. There is not, to be sure, anything like a proletarian class consciousness; nor a self–appointed revolutionary vanguard to lead the oppressed masses to freedom. Nor is it at all clear that American capitalism may not yet prove resilient enough to hold together indefinitely. But tensions of the sort Marxists call contradictions are intensifying in the capitalist system at a rate that should be of concern to defenders of free enterprise. In the long run, they may represent a more insidious threat to America's social well–being than ever did the Soviet military.

The Armed Forces' Current Role

What is the role of the armed forces in this period? The world over, it is business as usual: the business, in part, of defending national boundaries, but often that of aggressing against other peoples, or of oppressing the very people they are ostensibly defending, the citizenry of their own country. The particular mix of these functions varies from region to region and time to time. Where there are few external threats, as in Latin America, control of a country's citizenry often predominates. Where there are external threats, real or imagined, national defense tends to predominate. Even nations that do not use their militaries directly to oppress other peoples often back antidemocratic regimes whose military do this.

This has long been true in Latin America, but it is now particularly true in the Middle East, none of whose beneficiaries of massive U.S. and Soviet arms sales—with the notable exception of Israel (and Israel only for a portion of the population it controls)—is a democracy. U.S. policies there, once designed to counteract Soviet influence, now serve more to preserve its position as the dominant power of the region than to encourage democratization. Indeed, the movement toward democracy, particularly in countries like Algeria and Jordan, and the attempt to maintain the status quo in countries like Egypt, increasingly risks bringing to power fundamentalist Islamic movements with a decidedly anti–Western orientation.

Here we see the inherent incompatibility between WO1 and WO2.[11] WO1 would primarily ensure peace through collective security. It may be interventionist, on the grounds that national sovereignty immunizes governments against anything other than moral reproach for their treatment of their own citizens. But it may also be expressly interventionist, an understanding of it that is gaining acceptance today. Intervention may be thought to be warranted (and perhaps even mandated) when necessary to protect human rights.[12] According to WO2, on the other hand, the preservation of an order congenial to perceived U.S. (and perhaps Western) interests is the dominant concern. When democratization is thought to threaten those interests— as it did in Chile under the Allende regime, does in parts of the Middle East today, and may come to in Peru—it is democratization which suffers. Kuwait, Saudi Arabia, Egypt, and the Israeli occupied territories are all cases in point.[13]

America's Departure from Traditional Military Missions

In the U.S. there is increasing use of the military to deal with non-military problems, a change from its traditional role.[14] The sending of 27,000 troops into Panama to apprehend a head of state wanted on U.S. drug charges is only a particularly dramatic example.[15] The militarization of the so–called "war on drugs" in general symbolizes this change. It shows how tensions can be created even within this approach itself. For not only has the U.S. begun to depend increasingly upon the military to deal with the drug problem, it has encouraged others to do so as well. In Peru, U.S.–encouraged efforts to militarize the drug problem led to the alliance of many in the military with coca producers.[16] The beneficiary of millions of dollars in U.S. support and in–kind assistance in training, Peru's military has backed the suspension of democratic processes there, a development that did not go unnoticed in Bolivia, where there are concerns about a possible military coup.

In these cases, the military (in the first instance, that of the U.S. itself, in the second, that of other countries supported and trained by the U.S. military) are employed to use or threaten force directly in pursuit of the desired objectives.

But other roles are envisioned for the military—or for the war system in general. As General Carl E. Vuono wrote, when serving as the Army chief of staff:

> The United State provides military training, in one form or another, to 75 percent of the world's armed forces. This training is crucial to the successful assimilation of new weapons and tactics by friendly forces. More important, U.S. military training is a unique medium for encouraging the adoption of the values of professionalism, respect for human rights and support for democratic institutions.... Conventional forces, particularly the U.S. Army, actively support nation–building in countries throughout the world, assisting in the development of infrastructure that in turn, helps alleviate some of the root causes of instability and violence.[17]

One might add to this the 1991 relief operations on behalf of the Kurds in northern Iraq and Turkey, on behalf of flood victims in Bangladesh, and on behalf on Haitian refugees following the overthrow of the Aristide government, as well as efforts to save an Italian village from the lava flow of Mount Etna.[18] Vuono also points out the Army provided disaster relief following the California earthquake and Hurricane Hugo, fought forest fires in the West and supported antidrug campaigns on the border with Mexico. Both Army and Marine units were called in during the Los Angeles riots in 1992. What amounts to an institutionalization of such uses is contained in a Civil–Military Cooperative Action Program, proposed by Sam Nunn, chairman of the Senate Armed Services Committee, committing the military, at least during peace time, to a wide range of domestic activities, from rebuilding schools and bridges to providing a military–based youth training corps.[19] When one notes further the deep–rootedness of ROTC in American colleges and universities; the appointment of retired military personnel to the boards of directors of corporations; and, more recently, their hiring in increasing numbers for teaching positions in public schools, it is clear that, despite a reduction in military spending, there is an increase in the projected use of the military in the country's dealings abroad, and a growing role for it, and those trained in its values, in traditionally nonmilitary domestic institutions.

We are witnessing in the U.S. an almost imperceptible assimilation of military values to the approach to social, political, and moral problems, both domestically and internationally. This is happening at a time of decay in the country's social fabric. The educational system is deteriorating, poverty is increasing and punishment rather than social reform is looked to more and more as a solution to domestic problems. This is not "declinism" in a pejorative sense;[20] it is a realistic appraisal of tendencies inherent in the current course of events.

This social and economic decline is occurring at a time when the U.S. enjoys unrivaled global military power. But, if I am correct, that power, while it almost certainly will remain superior to any other for a long time to come, will likely be challenged eventually by Japan and Germany, as well as China. Add to this the growing signs that America's international prestige is declining as well, and the mix is volatile.[21] The temptation will be to try to compensate for our deteriorating social and economic situation by the exercise of military power to try to solve international problems in the way we increasingly try to solve domestic problems, by threats and punishment. What lies at the end of this road is not the world order dreamed of by neo–imperialists, but a world of increasing instability as the surviving superpower lashes out militarily at real or imagined threats in the attempt to stave off a more drawn–out version of the decline of its erstwhile adversary.

I emphasize that I am speaking only of tendencies here, for none of this is foregone. It will be the choices made through this transitional period that determine whether or not it comes to pass.

These choices, I suggest, need to reflect a redirection of our course as a nation, starting with a reconceptualization of the very idea of security. For just as social problems cannot in the long run be solved by SWAT teams and prisons, so security cannot in the long run be preserved solely or even primarily by military means. Even if we could keep our borders inviolate indefinitely, and could project sufficient power globally to be able to prevail in conflicts anywhere in the world, to do so would jeopardize the very values we would preserve. It would require becoming a virtual garrison state, geared forever to intervention abroad and chained to a permanent war economy at home.[22] This most likely would not happen by design, but gradually, almost imperceptibly, through prolonged breathing of the air of militarism, deceptively scented by the language of democratic values. A hint of this is found in Vuono's assertion that U.S. military training is a "unique medium" for encouraging respect for human rights and democratic institutions. Stripped of its glamour, and in plain language, military training aims in the first instance to produce specialists in the infliction of death and destruction—men and women who will perform efficiently, effectively, and on command. It is not a grand job–training program, or a prolonged summer camp, or a college preparatory school, as a surprising number of young people—influenced by aggressive advertising and recruiting—seem to view it. Those who undergo such training prepare either to engage in such killing and destruction themselves or to support (or in some cases, command) those who do. In so doing they must overcome, or help others to overcome, the natural inhibitions toward killing and causing harm—particularly on a large scale and to those they do not know and by whom they have never been personally wronged. The rest of what they do is incidental, whatever other functions it may serve. Seriously to encourage respect for human rights and democracy would be to support institutions designed to promote those values; to explore the historical, ethical, and political dimensions of the issues they raise in a spirit of open, critical inquiry. To expect respect for such values somehow to materialize as a by–product of the operations of an vast, bureaucratic institution that by the very nature of its mission is undemocratic and authoritarian, and that stands only a command away from being used in ways antithetical to those values, is wishful thinking.

Security and How to Get It

Security exists only when the conditions of a good life are present and sustainable. Physical survival is one of those conditions, but only one. An obsessive preoccupation with it, as one sees in so many nations, not only cannot guarantee the realization of the other conditions, it can actually obstruct their attainment; as it does in nations that squander their resources in an endless pursuit of physical security through armaments to the detriment, and sometimes impoverishment, of their own people, and often of those around them as well. Absolute security is an illusion, no more attainable by people collectively than by individuals. Acceptance of that fact is as essential for the well–being of a society as for the mental health of an individual.

The promotion and maintenance of the conditions of a good life, among them, notably, social justice, trust and respect for persons, requires cooperation, not only among members of one's own society, but among those persons and the members of other societies as well. This, in turn, requires a regard for the well–being of other peoples as well as one's own. In the end, the well–being of all peoples is interconnected; not in precisely the same way, or to the same degree, but in such a manner that no peoples' lasting security can be achieved at the expense of others. The exploitation, domination, or oppression of others eventually victimizes the victimizer as well, if only through the moral corruption it works in the oppressor's national character.

We think of military power as the best guarantor of security. But the collapse of the Soviet empire reveals the fragility of power understood principally as the capacity to inflict violence. A militarized state that could have been overcome, if at all, only at a horrendous cost if confronted head–on by military might, disintegrated once the true source of its power—the willingness of its people to continue acquiescing in the rule of a government in which they had long lost confidence—was removed. That happened once people lost their fear, as they did, not only in the Soviet Union, but also in the essentially nonviolent revolutions of Eastern Europe: from the ten-year struggle of Solidarity in Poland (whose members point with pride to the fact that they did not so much as break a single window during that time) to the six-week inspired uprising in Czechoslovakia. What they achieved almost certainly could not have been achieved through violence.

Lithuania is a case in point. The first of the Soviet Republics to declare its independence on March 11, 1990, Lithuania, under defense minister Audrius Butkevicius, mobilized a system of nonviolent defense against the Soviet military. Volunteers from throughout the country converged on Vilnius, setting up shacks in the newly christened Independence Square, site of the parliament building, and establishing communications networks. Lithuanians responded by the thousands when called out to confront Soviet troops nonviolently when they moved against key governmental and communications centers.[23] To be sure, the Soviets could have taken the parliament building—the seat of the new government, and symbolic of Lithuanian defiance of Moscow—in a few hours had they chosen to; and had not the Soviet government collapsed when it did, the Lithuanian struggle for independence would have been a protracted one of less certain outcome. But these considerations only highlight the fact that any successful nonviolent action, like any successful military action, presupposes conditions in whose absence the action might not have succeeded. Nonviolence did not succeed in Tiananmen Square. It did in Lithuania and throughout much of Eastern Europe. What is particularly significant about Lithuania, however, is the fact that nonviolence was deliberately incorporated into the government's defense strategy; indeed, it became its paramount—though never its sole—dimension. At no time, has a country expressly committed itself wholly to nonviolence. And only a very few (today, notably, Scandinavian countries) make nonviolence a component of their defense strategy. Nevertheless, it is in this direction, I want to suggest, that creative thinking must go in the quest for an alternative conception of security.

Pragmatic and Principled Nonviolence

Nonviolence can be thought of exclusively in practical terms, as simply another instrument of power for the attainment of social and political ends, or as a moral or spiritual philosophy—even a way of life—which may, though need not, be held to have broad social and political uses. The first conception may be called pragmatic, the second principled. Pragmatically, nonviolence is a form of power, differing from violent and violence–threatening forms but capable of serving the same kinds of ends, whether just or unjust. Disavowing as it does an essential moral or religious orientation, pragmatic non–violence is open to anyone to accept. It does not require superhuman, or even extraordinary, capacities—say, in the manner of a Gandhi or a Martin Luther King, Jr. The use of nonviolence particularly relevant to the issues under consideration here—that is, nonviolent national defense (or Civilian–Based Defense, as it is sometimes called)—requires only training and discipline on the part of ordinary citizens.

Principled nonviolence, on the other hand, though often viewed as a practical alternative to violence, and as requiring training and discipline, is taken to have a moral and/or spiritual foundation, usually requiring something of a transformation of its adherents. Nonviolence, on this view, is not simply a tactic or tool, to be dispensed with if it does not work. It has a different rationale. Living nonviolently, or being a nonviolent person, may be the principal objective. And it is one that may be attainable even in the face of failure to achieve more usual social or political objectives.

So there is a twofold difference between principled and pragmatic nonviolence: first, in the conception of what constitutes the justification of nonviolence, specifically whether it is moral (or spiritual) or purely practical, and second, in what a commitment to nonviolence is thought to entail—whether it requires leading a consistently nonviolent life or merely acquiring the necessary training and discipline to engage in nonviolent action where appropriate.

But whether principled or pragmatic, nonviolence does not threaten the lives or well–being of those against whom it is used; hence it minimizes the risk of at least such violence as is borne of fear for personal safety. And it is usable by virtually a whole population. Unlike military defense, which singles out one group (normally young men, though in this country increasingly young women as well) to bear the full burden of that defense, nonviolent national defense distributes responsibility throughout society. Not only can this give ordinary persons a stake in the defense of what they value, it can help alter the situation in which older men send boys and young men to fight without risk to themselves—a situation that does little to discourage military adventurism by heads of state.

There is not space here to detail the many techniques of nonviolent struggle—more that 180 of them have been catalogued by Gene Sharp.[24] But their aim, when used to counteract aggression, particularly an attempted takeover of one country by another, is the same as one of the stated objectives of military action: to deny to an adversary the ability to achieve his objectives. Rather than seeking to do this through force, however, nonviolence seeks to do it chiefly by withholding the cooperation necessary for the attainment of those objectives. In so doing it operate with a broader than usual understanding of power.[25] For political power involves more than merely the capacity to apply physical force. It requires the acquiescence of those over whom such power is held, and a willingness on their part to cooperate with the enterprises of those who govern them. Once that willingness ceases and cooperation is withdrawn, that power evaporates. There remains, of course, the capacity to inflict violence, to kill and destroy. But that alone rarely suffices to achieve one's objectives. And even that power presupposes a continuing willingness on the part of the military to carry out orders, and, at least in the long run, on the part of the citizenry of the offending country to support its government's aggression. All of these represent points of potential vulnerability that can be exploited by nonviolence, which seeks both to confront an aggressor's actions and to undermine his legitimacy.[26]

Choosing between Violence and Nonviolence

Does nonviolence work? As with military action, sometimes and sometimes not. There is no more assurance (at least when nonviolence is pragmatically conceived), than with violence, that any particular nonviolent campaign will succeed. That depends upon circumstances, including the training, discipline and morale of the practitioners of nonviolence, as well as upon that of their adversaries. But nonviolence has worked at specific times in the past—against the British in India, the French and Belgians in 1923, the Nazis in Norway and Denmark during World War II, and, of course, in the civil rights movement in the U.S. in the 1960s. And (though neither of these represents a clear commitment to nonviolence) it has played a significant role in the Palestinian Intifada, most notably in the civil disobedience against Israeli authorities in the tax resistance of the West Bank town of Beit Sahour, and is playing a role in the struggle of the Albanian majority against Serbian domination in the Yugoslavian province of Kosovo. Nonviolence does not promise simple solutions. It cannot guarantee there will be no bloodshed. It cannot, and does not seek to, end conflict. It holds open, rather, the promise of dealing creatively with conflict, resolving it where possible, managing it where not. It seeks to channel conflict constructively where its elimination would be undesirable. It holds open the promise, on the large scale, of providing an empowering, effective and less costly means of national defense.

But the question is not really whether nonviolence works. It is whether it works as well as, or better than, violence. For the choice is between those two approaches, and the shortcomings of either are relatively unimportant if, by comparison, those of its alternative are even greater. If nonviolence is relatively untried, violence is not, however, and its record is less impressive than is often supposed.[27] Every war that has a winner also has a loser, so that in every case in which violence succeeds on the victorious side, it fails on the losing side—an exact balancing of the scales.[28] Even the military success of Operation Desert Storm is offset by the failure of violence in the initial Iraqi aggression against Kuwait. Interestingly, the threat of force, which was represented as the surest way to peace during the military buildup of the fall of 1990, failed to dislodge Iraq from Kuwait; a fact that represented a failure of policy based on that threat when the 15 January 1991 deadline for Iraqi withdrawal passed. Even the military rout of Iraq, as effective as it was in getting Iraq out of Kuwait, did not achieve the clearly intended, though unstated, objective of bringing down Saddam Hussein. And it failed to achieve what came after the war to be represented as one of the war's principal aims, to destroy Iraq's nuclear capability. Iraqi compliance with U.N. resolutions in this regard, such as it has been, may well have owed more, in the end, to the effects of the embargo—the nonviolent sanction that was given less than a week to work before military measures were undertaken in August of 1990—than to the threat of force. And if one views the matter from the other side, Iraq has arguably been more effective in resisting U.N. pressures since the end of the war by what has, in effect, been a program of passive resistance (delay, indirection, noncooperation, tactical capitulations) than ever it was in its use of military force.

Both WO1 and WO2, and their many variants, as different as they appear on the surface, perpetuate the reliance upon force to get along in the world. In that sense there is nothing new about them. The most they can hope to achieve is a peace built upon violence and fear.

Building Institutions for Nonviolent Conflict Resolution

There is an unprecedented opportunity for the U.S., now that it is finally free of the Soviet threat, actively to promote the creation of institutions and mechanisms for nonviolent conflict resolution throughout the world.[29] This would hold some promise of creating a new world order in substance as well as name. Instead, it is pouring yet more arms into the hands of those who represent the status quo, as the militarization of the Middle East and Latin America, in particular, continues and that of South Asia gets underway in earnest.[30] This not only will not, but cannot, I suggest, bring about a new world order worthy of the name. The essential character of any world order is determined by the principal means relied upon to maintain it. And the means now envisioned in WO1 and WO2 are those of every failed order of the past— the use and threat of violence.

As long as the emphasis is primarily upon ends—even noble ends like freedom, democracy, and respect for persons—the way remains open for continuing reliance upon means that eventually subvert efforts to realize those ends. We need to shift our emphasis away from ends to means; to commit ourselves to nonviolent means that respect all persons equally, whether they happen to be Americans or not; and to limit ourselves in the pursuit of ends to those that are consistent with such means. This represents a reversal of the usual understanding of means and ends. Rather than adopting ends and then devising means to their attainment (which, on the international scene, invariably involves ultimately relying upon the instruments of violence), it counsels adopting means, specifically nonviolent means, and then pursuing only such ends as accord with them. Rather than the ends justifying the means, it is the means—at least to a significant degree—that justify the ends. This gives recognition to the fact, stressed by Gandhi, King, and others, that, at least insofar as we are viewing them from a moral perspective, means and ends cannot be separated from one another.

Commitment to promote a nonviolent world order would in this country, as well as in many others, require conversion to a peace–oriented rather than a war–oriented economy. Worldwide it would over time almost certainly entail a transformation of the nation–state system. For the modern nation–state gives social and political embodiment in today's world to centuries of humankind's habituation, through custom and convention, to relying upon the sword to achieve its ends.

Events in the former Soviet Union and parts of Eastern Europe are instructive in this regard. Many peoples there are trying to replicate on a smaller scale the same coercive state system they previously found oppressive as minorities within larger states; and doing so often with little regard for other ethnic groups that suddenly find themselves minorities in these newly established states (or, where they were minorities to begin with, finding themselves minorities under governments controlled by what had been another minority alongside them before independence). Ironically, they are in many cases insisting upon, and in the worst cases fighting over, boundaries established by the very regimes they now repudiate. In the overall process in which they are engaged, they are inadvertently laying bare the full nature of the institutionalization of violence represented by the state. It is as though, in these events, the world is witnessing in concentrated form, and at accelerated speed, the way in which most nations have been formed over the centuries: through forcible domination of some people by others.

But for the bloodshed to which it is leading, the dissolution of the Soviet Union would be a blessing to the world—as (all other things being equal) would be that of the world's other major military powers as well. The power to destroy countless persons at will, and perhaps even civilization itself, cannot safely be centralized in a handful of men or women, however chosen, and however seemingly trustworthy. Nation–states are not part of the nature of things. They certainly are not sacrosanct. If they perpetuate ways of thinking that foster division and enmity among peoples, ways should be sought to transcend them. Such ways are beginning to evolve now in any event, with the development of a global economy and increased travel and communications. The challenge is to give the process conscious, thoughtful direction.

There is the moral dimension to this, of course, in that if there should be a way for peoples and societies to get along in the world without depending upon military force, it behooves us to find it. It would represent an accomplishment for humankind surpassing any of science and technology. But moral considerations aside, national interest (or a public interest that might take its place with the transformation of the nation–state system) would call for finding it. We are on a course now in which the very fabric of society shows signs of coming unraveled through saturation with violence—not only as found in the streets, and increasingly in the schools, but as depicted in television, film, newspapers, popular music and culture. For this cannot help but take a toll on our greatest resource, our children. Not that conversion to nonviolent national defense would magically solve all domestic problems, including that of violence. But it would institutionalize a different set of values for dealing with international problems, setting the stage for the possibly even more difficult problem of cultivating methods of nonviolent conflict resolution for domestic problems as well. We cannot be a Rambo on the world scene, then turn around and tell young people to be nonviolent.[31] They will be what we as a people and a country are.

If the transition to a nonviolent world order should be possible (and it may not be), it is possible that ignorance and narrowness of vision are so deeply rooted that it cannot be done; I do not believe this, but it is possible—if, I say, such a transition should be possible, then at the very least discussions of strategy must begin to take seriously the possible role of nonviolence, initially, at least, as a component of current military thinking. Indeed, some conceptions of strategy already suggest as much (not by design, so far as I know, but in effect). As in one recent statement: "Today, strategy is not confined simply to devising ways to win battles; it is a daily preoccupation of most modern governments—certainly most industrial states—and its scope has broadened to include all of the elements of national power, not just the military."[32] Taking this at face value, the potential of nonviolence, and its possible effectiveness in national or social defense, must be reckoned among the elements of national power,[33] and hence taken into account in strategy, even from a position of *realpolitik*. For the effectiveness of nonviolence—leaving moral considerations aside—is an empirical issue. It can be assessed only by the same rigorous scrutiny and testing commonly brought to bear upon alternative military strategies; it cannot be assessed a priori.

If the vast material and human resources we now expend in the quest for security through military means are not affording us that security, and if that effort is costing us more and more in our ability to deal constructively with challenges to the values we profess to cherish, realism in any meaningful sense argues for taking up the challenge of nonviolence.

Notes

1. *Washington Post National Weekly Edition*, May 27–June 2, 1991.
2. See Richard Rosecrance, "A New Concert of Powers," *Foreign Affairs* 71 (Spring 1992), 64–82.
3. See Joseph S. Nye, Jr, "What New World Order?" *Foreign Affairs* 71 (Spring 1992), 83–96.
4. See, for example, Ted Galen Carpenter, "The New World Disorder," *Foreign Policy* 84 (Fall 1991), 24–39.
5. A report on a draft of the document was published in *New York Times*, March 8, 1992.
6. See *New York Times*, May 24, 1992.
7. He also saw this conception as lamentably lacking in realism. See George F. Kennan, *American Diplomacy: 1900–1950* (New York: New American Library, 1951), 82.
8. I have stated both in pure forms, ignoring the complexities and permutations of each and the extent to which there might be a mix of the two.
9. The German military sent mine sweepers to the Gulf waters following the Gulf War, and has contributed helicopters to U.N. observer units in Iraq. In addition to aiding the Kurds in northern Iraq, it helps operate a U.N. hospital in Cambodia. On these issues, see Francine S. Kiefer, "Germany Tiptoes Toward Greater Use of Military," *Christian Science Monitor*, July 16, 1992.
10. See, for example, John J. Mearsheimer, "Back to the Future: Instability in Europe After the Cold War," *International Security* 15 (Summer 1990).
11. Despite the fact that some, like Joseph Nye, Jr., seem to think that effecting a blend of the two will best serve U.S. interests. See his "What New World Order?"
12. For a good recent statement of the case for human rights based interventionism on internationalist grounds, see Fernando R. Teson, *Humanitarian Intervention: An Inquiry Into Law And Morality* (Dobbs Ferry, N.Y.: Transnational Publishers, 1988).
13. This readiness to see democratic aspirations stifled may coexist, of course, with a willingness to shake a finger at human rights violations and the undermining of democracy, as it does in Haiti and Peru. But that coexistence, even when it is not the intention, will likely serve mainly to preserve the appearance of WO1 all the while the forces of WO2 are at work.
14. Though not altogether without precedent, as the Army Corps of Engineers has long been involved in civilian projects, and the Civilian Conservation Corps was run by the military.
15. This was not the sole purpose, which included protecting American lives and safeguarding the canal.
16. As it has also in Colombia, as Colombian lawyer and human rights advocate Jorge Gomez Lizarazo writes in an op–ed piece: "U.S. aid, justified in the name of the drug war, is furthering the corruption of the Colombian security forces and strengthening the alliance of blood between right–wing politicians, military officers and ruthless narcotics traffickers." *New York Times*, January 28, 1992.
17. Carl E. Vuono, "Desert Storm and the Future of Conventional Forces," *Foreign Affairs* 70 (Spring 1991), 55.
18. See Eric Schmitt, "U.S. Forces Find Work As Angels of Mercy," *New York Times*, January 12, 1992; Scott Shuger, "Pacify the Military" *New York Times*, March 14, 1992; and Alan Cowell, "It's Plug Up Mt. Etna or Go the Way of Pompeii," *New York Times International*, April 25, 1992.

19. Eric Schmitt, "Civilian Mission Is Proposed for Post–Cold War Military," *New York Times*, June 24, 1992. See also Scott Shuger, "Pacify the Military," *New York Times*, April 14, 1992.

20. As characterized by Samuel P. Huntington, in "No Exit: the Errors of Endism," *National Interest* (Fall 1989): 3–11.

21. As John Lukacs writes, "The power of a nation, like that of a person, is inseparable from the unquantifiable asset of its prestige." John Lukacs, "The Stirrings of History: A New World Rises from the Ruins of Empire," *Harper's Magazine*, August 1990.

22. I borrow this term from Seymour Melman, *The Permanent War Economy* (New York: Simon and Schuster, 1974).

23. The most dramatic event was the move by Soviet tanks against the Vilnius television tower, only to find thousands of Lithuanians converging on the tower, which they surrounded in nonviolent defense. The Soviets killed thirteen people in the course of occupying the tower, but this left them in a position of moral vulnerability, as they sat, surrounded by fence and barbed wire, peered at by Lithuanians gathering to lay wreaths honoring the dead.

24. According to eyewitness accounts, Sharp's list of nonviolent techniques was publicly posted near the Russian White House early in the August 1991 coup attempt.

25. As Gene Sharp says, "Civilian–based defense rests on the theory that political power, whether of domestic or foreign origin, is derived from sources *within* each society. By denying or severing these sources of power, populations can control rulers and defeat foreign aggressors." *Civilian–Based Defense: A Post–Military Weapons System* (Princeton, N.J.: Princeton University Press, 1990), 7.

26. As may the use of so–called "indirect force" as understood by Sun Tzu as well. As Max G. Manwaring observes, in characterizing such force as an adjunct to military force: "Indirect force is applied or conducted through the use of moral power. If carefully done, the use of moral influence can undermine the legitimacy and the position of another actor by breaking the bonds which unite a people, its political leadership, and its protective military/police organization. When legitimacy is seriously questioned, will is destroyed and the opponent is weakened to the point where only a minimum of direct military force is necessary to assure the desired change. Moreover, by transforming the emphasis of war from the level of military violence to the level of a struggle for legitimacy, an actor can strive for total objectives (e.g., the overthrow of a government) instead of simply attempting to obtain leverage and influence for 'better terms' in the classical dimension." "Limited War and Conflict Control," in Stephen J. Cimbala and Keith A. Dunn, eds., *Conflict Termination and Military Strategy: Coercion, Persuasion, and War* (Boulder, Colo.: Westview Press, 1987), 60.

27. John Mueller, in a critical assessment of the efficacy of the nuclear threat, argues in fact that: "It seems likely that the vast majority of wars that never take place are caused by factors which have little to do with military considerations." See his "The Essential Irrelevance of Nuclear Weapons," *International Security* 13 (Fall 1988), 18.

28. For a discussion of this issue, see the General Introduction to my *Non–violence in Theory and Practice* (Belmont, Cal.: Wadsworth, 1990).

29. One of the few public calls for this is found in Helena Cobban, of Search for Common Ground in Washington, who writes: "We should actively promote, worldwide, the work of those American individuals and institutions that over past decades have pioneered nonviolent ways to resolve inter–group conflicts.... These institutions have put the U.S. at the cutting–edge of the discipline of con-

flict resolution. This kind of "software," rather than further sales of military hardware, is what we ought to be offering a world facing new insecurities at every turn." *Christian Science Monitor*, March 19, 1992.

30. With the U.S. having sold more $14.8 billion in arms to Saudi Arabia, $2.17 to Egypt, $737 million to the UAR, and $3.2 billion to South Asia during the year following the Gulf War. James Adams, *Washington Post National Weekly Edition*, March 23–29, 1992. During 1990, the U.S. became the world's biggest arms supplier overall, accounting for 40 percent of the market. Ira Shorr, "Making History in a Post Cold War World, *Global Security News*, A periodic publication of the SANE/FREEZE Education Fund, Winter 1991–92.

31. As one sixteen-year-old girl reportedly said with exasperation to then presidential hopeful, Governor Bill Clinton, at a New York high school, "I mean, we went to war over oil, and we're telling our kids not to shoot each other?" *New York Times*, March 25, 1992.

32. Keith A. Dunn and William O. Staudenmaier, "US Military Strategy in the Nuclear Era," in Keith A. Dunn and William O. Staudenmaier, eds., *Alternative Military Strategies for the Future* (Boulder, Colo.: Westview Press, 1985), 213.

33. In a glimmering of recognition of this shortly before the collapse of the Soviet Union, John Lewis Gaddis wrote: "What seems most likely is not that some new rival will emerge, capable of challenging the superpowers militarily, but rather that the standards by which we measure power will begin to evolve, with forms other than military—economic, technological, cultural, even religious—becoming more important." John Lewis Gaddis, "How the Cold War Might End," *The Atlantic Monthly* (November 1987).

Contributors

James Burk is professor of sociology at Texas A&M University and was formerly editor of *Armed Forces and Society*. He is author of many articles on civil–military relations and editor of *Morris Janowitz: On Social Organization and Social Control*.

Christopher Dandeker is professor and head of the Department of War Studies, King's College, University of London and co–founder of the British Military Studies Group. He is author of *Surveillance, Power, and Modernity* and numerous articles on historical trends in military organization.

Robert L. Holmes is professor of philosophy at the University of Rochester. He is author of *On War and Morality* and editor of *Nonviolence in Theory and Practice*.

Charles C. Moskos is professor of sociology at Northwestern University and formerly chair of the Inter–University Seminar on Armed Forces and Society. He is author of *A Call to Civic Service* and co–author of *All That We Can Be*.

James N. Rosenau is university professor of international affairs at George Washington University. He is author of *Turbulence in World Politics* and *Along the Domestic–Frontier: Exploring Governance in a Turbulent World*.

David R. Segal is professor of sociology and government at the University of Maryland, chair of the Inter–University Seminar on Armed Forces and Society, and was formerly editor of *Armed Forces and Society*. He is author of *Recruiting for Uncle Sam* and co–author of *Peacekeepers and Their Wives*.

Donald M. Snow is professor of political science at the University of Alabama. He is author of *The Shape of the Future: The Post–Cold War*

World, From Lexington to Desert Storm, and *Distant Thunder: Patterns of Conflict in the Developing World.*

William R. Thompson is professor of political science and director of the Center for the Study of International Relations at Indiana University. He is co-editor of *International Studies Quarterly* and co-author of *The Great Powers and Global Struggle, 1490–1990, War and State Making, On Global War*, and *Leading Sectors and World Powers.*

Robert J. Waldman is with the U.S. Public Health Service in Washington, D.C. He completed his Ph.D. in sociology at the University of Maryland in 1991.

Index